Julian Antoni Skolaut

Molecular Motor Based on Single Chiral Tripodal Molecules Studied with STM

Experimental Condensed Matter Physics
Band 31

Herausgeber
Physikalisches Institut
Prof. Dr. David Hunger
Prof. Dr. Alexey Ustinov
TT-Prof. Dr. Philip Willke
Prof. Dr. Wolfgang Wernsdorfer
Prof. Dr. Wulf Wulfhekel

Eine Übersicht aller bisher in dieser Schriftenreihe erschienenen Bände finden Sie am Ende des Buchs.

Molecular Motor Based on Single Chiral Tripodal Molecules Studied with STM

by
Julian Antoni Skolaut

KIT Scientific Publishing

Karlsruher Institut für Technologie
Physikalisches Institut

Molecular Motor Based on Single Chiral Tripodal Molecules
Studied with STM

Zur Erlangung des akademischen Grades eines Doktors der Naturwissenschaften von der KIT-Fakultät für Physik des Karlsruher Instituts für Technologie (KIT) genehmigte Dissertation

von Julian Antoni Skolaut, M.Sc.

Tag der mündlichen Prüfung: 24. Juni 2022
Referent: Prof. Dr. Wulf Wulfhekel
Korreferent: Prof. Dr. Wolfgang Wernsdorfer

Impressum

KIT Scientific Publishing

Karlsruher Institut für Technologie (KIT)
KIT Scientific Publishing
Straße am Forum 2
D-76131 Karlsruhe

KIT Scientific Publishing is a registered trademark
of Karlsruhe Institute of Technology.
Reprint using the book cover is not allowed.

www.ksp.kit.edu

Print on Demand 2023 – Gedruckt auf FSC-zertifiziertem Papier

ISSN 2191-9925
ISBN 978-3-7315-1247-9
DOI 10.5445/KSP/1000151650

Contents

1 Introduction

For the past decades, digitalization has influenced major parts of society intimately in everyday life. With the rise of smartphones, it is possible today to hold devices in your hand with much more computing power than first computers, filling rooms back in the 1940s. This rapid progress has been driven by ever-decreasing sizes of basic electronic components, such as transistors, which is made possible by improving fabrication techniques of the semiconductor industry, such as lithographic processes. However, the pace of this progress is slowing down and, as components become smaller, traditional fabrication processes reach limitations of a fundamental nature. One such limitation is the diffraction limit of optical processes in lithography, but at the approached scales, also quantum mechanical effects gain significance. This has led to significant efforts in trying to realize electronic components on molecular scales, following a bottom-up, rather than a top-down approach.

These efforts are not only limited to electronic components. Inspired by the success of miniaturization in electronics, also other systems, such as nanomachines were envisioned. Starting from nanomachines on the scale of larger proteins, such as myosin that converts chemical energy to motion along the cytoskeleton of cells, researchers got more interested in smaller and smaller machines and began to artificially create molecules with the goal to turn external stimuli, including light and electrical current, into motion. In 2016 the nobel prize in chemistry was awarded to Bernard L. Feringa for the design and synthesis of molecular machines [1, 2].

This idea is also pursued in the investigations presented in this thesis, in which it is sought to implement a complete motor in a single, specifically synthesized molecule, driven by the tunneling current in a scanning tunneling microscope (STM). The synthesis of the tailor-made molecules presented in this thesis is done at the Institute of Nanotechnology (INT) at the Karlsruhe Institute of Technology (KIT), by Dr. Nico Balzer and Dr. Michal Valášek in the group of Prof. Marcel Mayor. To enable the usage of the molecules as a motor, they were deposited on a surface, here chosen to be gold, on which they are fixed through sulfur-gold bonds of a tripodal platform. This platform serves as the stator of the motor molecule. The other necessary parts of a motor, namely axle and rotator are implemented by mounting a functional group on the platform *via* a freely rotatable acetylene group, which, in combination with the platform, also decouples the rotator from the surface.

The functional group is chosen to be chiral, in an attempt to harvest the spin-polarization caused by chiral-induced spin selectivity (CISS) to transfer a torque on the functional group, defining the preferred rotation direction needed to call the molecule a motor.
As the CISS effect is intended to direct the rotation of the motor molecules, additional efforts investigate the effect in the form of conductive atomic force microscopy (CAFM) studies on self-assembled monolayers (SAMs) of chiral molecules on gold surfaces. They were acquired at Leiden University in the Netherlands, in the group of Prof. Jan M. van Ruitenbeek in the framework of a Research Travel Grant provided by the Karlsruhe House of Young Scientists (KHYS).
This thesis consists of experimental results acquired with two different techniques, namely STM and CAFM. After presenting fundamentals necessary for the interpretation of the experimental results in Chap. 2, the CAFM setup used for the studies on SAMs of chiral molecules is presented in Chap. 3.
The results obtained with this setup, discussed in Chap. 4, include studies on the inversion of CAFM probe magnetization needed to invert the spin-polarization of the injected current, studies on reference gold substrates and on SAMs of achiral molecules. The measurements on the reference samples emphasize the importance of the choice of substrate for CAFM studies of molecular SAMs. Finally, studies on chiral molecule SAMs show how challenging it is to obtain reliable results within CAFM studies on these molecules and conclude the investigation of molecular SAMs.
Chapter 5 presents the low-temperature STM setup used to study single molecules and ordered islands thereof, as well as their directed rotation. To achieve reliable observation of the directed rotation, Chap. 6 presents, how various molecules adsorb on the gold surface, finding that a combined deposition of two types of molecules with the same platform serves as a suitable system.
Being able to frequently measure directed rotation, Chap. 7 elucidates the dependences of various figures of merit of the rotation on experimental parameters, namely the lateral tip position, bias voltage and polarity and current. It reveals that the observability of the rotation depends strongly on the choice of tip–molecule distance, that the rotation is only asymmetric (directed) in a small voltage window and that the direction of rotation is not inverted for opposite bias polarity.
In addition to the experiments, calculations of the molecular structure and orbitals performed in the group of Prof. F. Evers are shown. Finally, possible underlying effects of the preferred rotation direction are checked with regard to their compatibility with both calculations and experimental results.
Chapter 8 gives a brief summary of the results presented in this thesis and discusses possible further experiments.

2 Theoretical Background

Fundamentals, which are necessary to understand the performed experiments, their analyses and interpretation comprise the contents of Chap. 2. Section 2.1 explains the working principle of the scanning tunneling microscope, namely quantum tunneling and its approximations for the specific case of scanning tunneling microscopy (STM). Section 2.2 discusses fundamentals of the experimental technique used to study self-assembled monolayers (SAMs) of molecules, namely conductive atomic force microscopy (CAFM). Section 2.4 presents possible drives for a molecular motor in the scope of this work. Finally, in Sec. 2.5, the theoretical principles used in the evaluation of the gathered data are discussed.

2.1 Scanning Tunneling Microscopy

In STM, an atomically sharp tip is placed in close vicinity (typically a few angstroms) of a conductive object. Applying a voltage to the two electrodes (tip and sample), the quantum mechanical tunneling effect leads to a tunneling current running through the junction. This current not only depends exponentially on the distance between the electrodes, but also varies with their electronic properties. In combination with the very precise positioning of the tip, STM allows for the investigation of topographic and electronic properties of materials with unmatched spatial resolution. The dependence of the tunneling current on the properties of the sample, the basic principles of tunneling and STM operation will be described in the next sections.

2.1.1 Quantum Tunneling and STM-Specific Approximations

In classical physics it is impossible for a particle to pass a potential barrier if it has insufficient energy. In quantum mechanics, however, the wave-like nature of electrons makes it possible which is called the tunneling effect. The most elementary way to describe this effect is to consider a particle tunneling through a barrier of fixed height in one dimension. In this case, the potential V can be described by

$$V(z) = \begin{cases} 0 & z < 0 \\ V_0 & 0 < z < a \\ 0 & z > a \end{cases}, \tag{2.1}$$

where V_0 is the finite barrier height. The wave function $\psi(z)$ of the particle can be derived by solving the stationary Schrödinger equation

$$E\psi(z) = -\frac{\hbar^2}{2m}\frac{\partial^2\psi(z)}{\partial z^2} + V\psi(z), \tag{2.2}$$

in which i is the imaginary number, m is the mass of the particle, $\hbar$ is the reduced Planck constant and E is the energy of the particle. Applying a plane wave ansatz for the three regions of the potential

$$\psi(z) = \begin{cases} A_1 e^{ikz} + A_2 e^{-ikz} & , z < 0 \\ A_3 e^{\kappa z} + A_4 e^{-\kappa z} & , 0 < z < a \\ A_5 e^{ikz} & , z > a \end{cases} \tag{2.3}$$

solves the Schrödinger equation with $k = \sqrt{2mE}/\hbar$ and $\kappa = \sqrt{2m(V_0 - E)}/\hbar$. For $E < V_0$, this leads to oscillatory behavior on both sides of the barrier and an

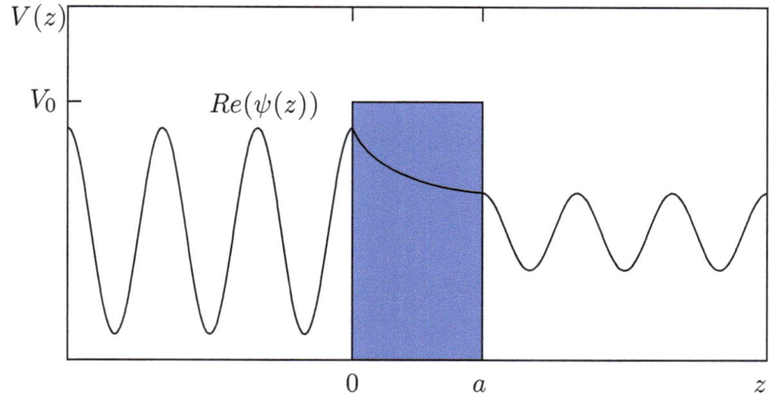

Figure 2.1: Sketch of 1D quantum tunneling. The blue shaded area is the barrier with height V_0 which is larger than the energy of the particle described by ψ. The wave function has a non-zero amplitude on both sides of the barrier. The particle can cross the barrier, even though it would not be allowed classically.

exponential decay within the barrier, as shown in Fig. 2.1. Even though classically, the particle does not have enough energy to cross the barrier, the probability to find it on both sides of the barrier $|\psi|^2$ is non-zero.

Taking into account the boundary conditions for the continuity of ψ and its derivative $d\psi/dz$ at $z = 0$ and $z = a$, the coefficients A_1 to A_5 can be calculated. This yields the transmission coefficient T for the particle through the barrier which is proportional to the tunneling current I and can be expressed as

$$I \propto T = \left|\frac{A_5}{A_1}\right|^2 = \frac{4E(V_0 - E)}{4E(V_0 - E) + V_0^2 \sinh^2(\kappa a)} \propto e^{-2\kappa a}, \tag{2.4}$$

meaning that the tunneling current depends exponentially on the width of the potential barrier. In a typical case in STM, the current increases by roughly an order of magnitude for a change of 1 Å in tip–sample distance. This behavior is one reason making STM such a powerful tool in the characterization of surfaces. It allows for a very precise determination of height profile of a sample.

Although this description provides a basic understanding of quantum tunneling, it does not fully capture the process. A more detailed description was provided by Bardeen in 1961 [3]. It was initially intended as a model for a solid state conductor–insulator–conductor junction, but it can also be applied to STM.

In contrast to the simplified model, it takes into account that the tunneling electrons are Fermions following Fermi statistics and can therefore tunnel only from an occupied into an unoccupied state. Their occupation probability at a certain temperature is given by the Fermi-Dirac distribution $f(E)$. Because the two electrodes are sufficiently far apart, they are described as separate metals and are assumed to possess non-interacting electrons with separate Fermi energies E_F^t and E_F^s for the

tip and sample, respectively. The applied bias voltage adds an energy offset of eU between the electrodes. Finally, the tunneling current is described in the framework of first-order perturbation theory.
For this, Fermi's golden rule is applied, which takes the form of

$$I = \frac{4\pi e}{\hbar} \sum_{\mu,\nu} [f(E_\mu) - f(E_\nu + eU)] |M|^2 \delta(E_\mu - E_\nu - eU) , \tag{2.5}$$

in which e is the electron charge and the indices μ and ν enumerate the tip and sample states, respectively. The transition matrix element is $M = \int_S \psi_\mu^* \nabla \psi_\nu - \psi_\nu^* \nabla \psi_\mu dS$ with S denoting a surface where the wave functions of tip and sample overlap. This expression can be conveniently re-expressed in terms of the local density of states (LDOS) ρ^t and ρ^s for tip and sample, respectively

$$I = \frac{4\pi e}{\hbar} \int \rho^t(E_F^t + E) \rho^s(E_F^s + E + eU) [f(E_F^t + E) - f(E_F^s + E + eU)] |M|^2 dE . \tag{2.6}$$

Equation 2.6 already shows that the tunneling current does not only depend on the tip–sample distance, but also on the LDOS of tip and sample.
Although expression 2.6 describes the problem more accurately, it is rather inconvenient for use, since the transition matrix element can exhibit a very complicated energy dependence. The problem was addressed by Tersoff and Hamann, who developed a STM-specific approximation [4] where the tip wave functions contributing to the tunneling current are assumed to be spherically symmetric s-wave functions. The tip is described as a sphere of radius R at position $\vec{r}_0$ resulting in a tip–sample distance of d. Assuming low temperatures and small applied bias voltages, this yields the expression

$$I = \frac{e}{\hbar} \int_0^{eU} \rho^s(\vec{r}_0, E_F^s + eU + E) \rho^t(E_F^t + E) T(E, U, d) dE , \tag{2.7}$$

with the transfer probability

$$T(E, U, d) = \exp\left(-\frac{2d}{\hbar} \sqrt{m(\Phi^s + \Phi^t + eU - 2E)} \right) , \tag{2.8}$$

where Φ^t and Φ^s denote the work functions of tip and sample, respectively. To further simplify the equation, it is often assumed that the LDOS of the tip is comparably flat close to the Fermi energy and that the transfer probability is not energy-dependent. This leaves only the LDOS of the sample under the integral. This fact is exploited in scanning tunneling spectroscopy (STS) measurements, where the LDOS of the sample is measured as the derivative of the tunneling current with respect to the bias voltage:

$$\frac{dI}{dU} \propto \rho^s(\vec{r}_0, E_F^s + eU) . \tag{2.9}$$

However, the tip cannot always be described by an s-wave function and, therefore, drastic deviations from the expected behavior in experiments can occur.
All processes described are based on elastic electron tunneling. However, it is also possible for the tunneling electrons to deposit part of their energy during the tunneling process and excite an inelastic excitation, as long as the electron energy (eU) exceeds the excitation energy ($\hbar\omega_e$). These excitations can be of various kinds, for example vibrational modes and phonons [5, 6], or spin-flip excitations [7, 8]. This opens an additional tunneling channel, increasing the tunneling probability and therefore the tunneling current. This can be observed in I/U spectroscopy as a pair of kinks, or as a pair of steps in dI/dU measurements, at the threshold voltage at positive and negative bias.
Scanning tunneling microscopy can be employed to study both the topography and the electronic structure of surfaces with extremely high precision in various ways, which will be described in the following.

2.1.2 Modes of Operation

In STM three basic modes of operation are distinguished, namely constant height mode, constant current mode and scanning tunneling spectroscopy (STS). As a prerequisite, a coordinate system has to be defined. Here, the (x,y) coordinate plane is chosen to be parallel and z is perpendicular to the sample surface.
In constant current mode, for a certain bias voltage, a setpoint of the current is fixed. The tip is moved laterally in fixed increments corresponding to pixels, forming an image of the surface. At every pixel, if there is a change in current, e.g. due to a change in overlap of the wavefunctions of tip and sample, the z position of the tip is adjusted so that the tunneling current matches the setpoint (see Fig. 2.2a), achieved with a feedback loop. Plotting the z-positions yields the surface of constant tunneling probability. If the LDOS is the same throughout the measurement, this corresponds to the topography of the sample. This measurement method is a comparably safe way to image a surface, since it avoids a collision of tip and sample.
In constant height mode, the tip is kept at a fixed z position and scans the desired area. At every pixel, the tunneling current is measured (see Fig. 2.2b). Since the tunneling current is very sensitive to the overlap of the wavefunctions of tip and sample, this mode is most suited to study the object closest to the tip apex. However, this sensitivity falls off very quickly with distance. So for the imaging of objects with high aspect ratios perpendicular to the sample surface, it is only suitable to image the topmost part of the object. For measurements in this mode it has to be clarified in advance that the scanned area shows no large corrugation, avoiding collisions of tip and sample.

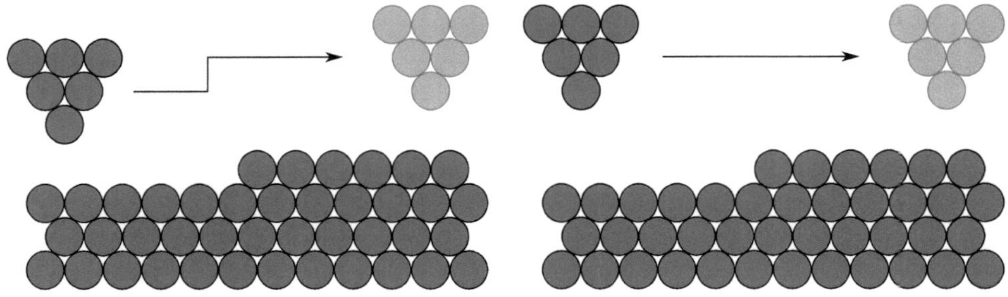

(a) Constant current mode. The z position is adjusted to maintain the current setpoint.

(b) Constant height mode. The z position of the tip is fixed and the current is measured.

Figure 2.2: Modes of operation for imaging in STM.

The last mode of operation is STS, where the lateral tip position is fixed, while one parameter (e.g. bias voltage or z position) is swept and a second parameter (e.g. current or differential conductance) is measured. It allows for measurements of the LDOS of the sample at a certain tip position, giving insight into its electronic structure, as outlined in Eq. 2.9.
All methods are well-suited to investigate molecules on surfaces. In the context of this thesis, the molecules are intended to function as motors. Section 2.3 presents a brief review of the history of studies of molecular motors. To support the studies on single molecular motors potentially driven by chiral-induced spin-selectivity (CISS), it is tried to study CISS directly using self-assembled monolayers of chiral molecules with conductive atomic force microscopy (CAFM). Therefore, Sec. 2.2 presents a brief introduction to atomic force microscopy and two variations thereof, namely CAFM and magnetic force microscopy.

2.2 Atomic Force Microscopy

Atomic force microscopy (AFM)[9] is a widely used technique for surface studies from the family of scanning probe microscopies (SPMs), as is STM. Similar to STM, a sharp tip, often called probe in the context of AFM (usually in the order of 10 nm apex radius, mounted on a micron-scale cantilever), is brought close to the surface and is moved in the (x,y) plane (same coordinate system as defined in 2.1.2) to gain information about the topography and other features of the surface (e.g. adhesion). However, in AFM, atomic forces between the probe and the surface are used to extract information on the sample, in contrast to the tunneling current in STM. The working principle of the AFM is described in more detail in Subsec. 2.2.1. Again, the scanning of the surface can be done in different modes (see Subsec. 2.2.2). To extract additional information on the sample, such as the conductivity or magnetic properties, variants of AFM are used, namely conductive AFM (CAFM, Subsec. 2.2.3) and magnetic force microscopy (MFM, Subsec. 2.2.4).

2.2.1 Working Principle

AFM uses atomic forces between a sharp tip (called probe) and a sample to study the sample surface. The exerted forces have two main contributions, the attractive Van-der-Waals (VdW) force and the repulsive force caused by the Pauli exclusion principle.

The VdW force describes a long-range attractive interaction due to induced dipoles in neighboring atoms. It is possible to spontaneously form a dipole, caused by the distribution of negatively charged electrons around the positively charged core. This dipole induces a dipole of opposite polarity in the neighboring atom, effectively causing an attractive interaction. It is usually described by a potential proportional to $\frac{1}{r_{\mathrm{LJ}}^6}$, where r_{LJ} is the atomic distance.

The short-ranged force caused by the Pauli exclusion principle is due to electrons in the shells of the atoms in contact, not being allowed to occupy the same states. Therefore, the electrons are pushed to unoccupied states at higher energies, which is unfavorable. The potential caused by the Pauli repulsion is often chosen to be proportional to $\frac{1}{r_{\mathrm{LJ}}^{12}}$ for convenience.

These contributions combined result in the so-called Lennard-Jones potential, which describes the forces between two atoms and reads

$$V_{\mathrm{LJ}}\left(r_{\mathrm{LJ}}\right) = 4\epsilon \left[\left(\frac{\sigma_{\mathrm{LJ}}}{r_{\mathrm{LJ}}} \right)^{12} - \left(\frac{\sigma_{\mathrm{LJ}}}{r_{\mathrm{LJ}}} \right)^{6} \right], \tag{2.10}$$

where $\epsilon > 0$ is a measure of the depth of the potential well caused by the VdW force and Pauli repulsion and σ_{LJ} is the atomic distance at which the potential crosses zero.

A sketch of the Lennard-Jones potential is shown in Fig. 2.3. At large distances, the force is small, increasing towards smaller distances due to the VdW force. It reaches maximally attractive potential at $r_{\mathrm{LJ},m}$, where the repulsive forces starts to become more significant. At $r_{\mathrm{LJ}} = \sigma_{\mathrm{LJ}}$, the repulsive force supersedes the attractive force.

As AFM is also an SPM technique, the schematic setup is very similar to STM and can be seen in Fig. 2.4. The probe is brought so close to the surface (with a piezo scanner), that they touch and a force is exerted on the probe mounted at the end of a

[1] `https://creativecommons.org/licenses/by/4.0/legalcode`

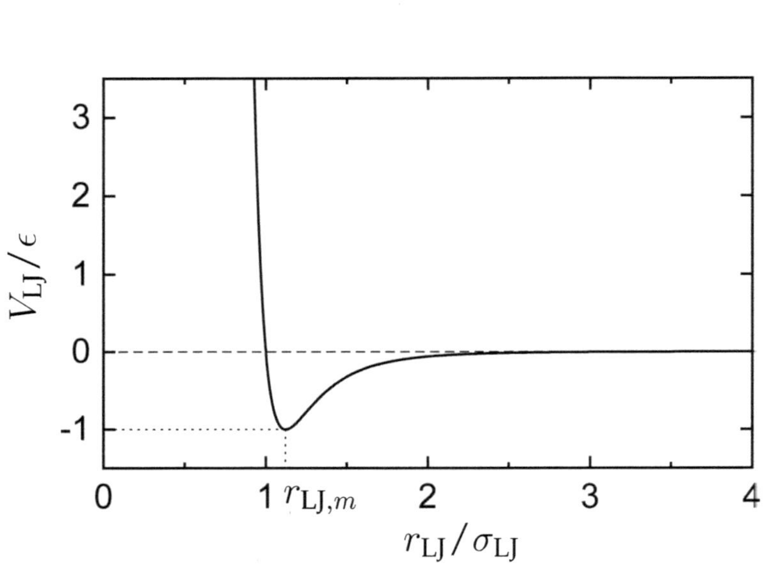

Figure 2.3: Sketch of the Lennard-Jones potential describing the forces acting between two atoms. At larger distances, it is attractive due to VdW forces and at small distances, it is repulsive due to the Pauli exclusion principle (adapted from [10] under CC BY 4.0 license[1]).

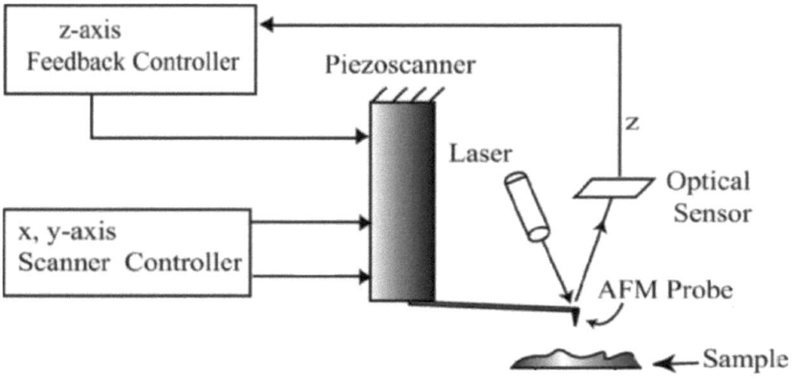

Figure 2.4: Schematic of the scanning principle of AFM. The cantilever bending is detected through the deflection of a laser beam, corresponding to a force. Scanning in the (x,y) plane, the z position is adjusted according to the deflection by a feedback controller (adapted from [11], ©2005 IEEE).

flexible cantilever. The force on the probe can be detected by various methods. If the backside of the cantilever is coated with metal, another electrode can be brought close to it. Applying a voltage to both sides, the distance can be detected by measuring the capacitance [12]. Using a piezoelectric material as as probe material, the deflection can directly be measured by the electrical signal [13]. However, the most commonly used technique is to detect the bending of the cantilever, proportional to the exerted force, by the deflection of a laser beam reflected from the back side of the cantilever (see Fig. 2.4).

To be able to convert the deflection of the beam, which corresponds to a deflection of the cantilever, to an applied force, a calibration of the cantilever is necessary. The force and displacement of the cantilever, which is treated as a spring, are connected *via* Hooke's law. The crucial part of the calibration process is to find the spring constant k_{cant}.

The most common technique is so-called thermal tuning, in which the cantilever's fundamental resonance is excited, based on the equation derived by Sader *et al.* for rectangular cantilevers [14]

$$k_{\text{cant}} = 0.1906 \rho_f b^2 L Q_f \Gamma_i(\omega_f) \omega_f^2 \,, \tag{2.11}$$

in which ρ_f is the density of the fluid in which the oscillation takes place, b and L are the width and length of the cantilever, respectively, Q_f is the quality factor of the fundamental mode in fluid, Γ_i is the imaginary part of the hydrodynamic function [15] and ω_f is the resonance frequency in fluid.

Most of the parameters needed for this calculation are known, such as the measures of the cantilever (b and L) and the properties of the fluid (ρ_f and Γ_i). Only the resonance frequency and quality factor have to be found. This can be done by exciting the fundamental resonance of the cantilever by sweeping the frequency of excitation and fitting a Lorentzian-type [16] function to the resonance. Therefore, the needed quantities can be obtained without a specific sample and without being in direct contact with the surface, making this method easy to use and readily available. In fact, it is often implemented as an automatic cantilever calibration procedure in AFM control software.

Knowing the connection between laser beam deflection and force, it can be held constant by a feedback system upon scanning point by point in the (x,y) plane. Adjusting the z position of the probe for every lateral position, the topography of the surface can be mapped.

2.2.2 Modes of Operation

As for the STM, there are several modes of operation of an AFM. Here, three commonly used modes (contact, intermittent contact (AC) and force spectroscopy), as well as two more specific modes (Hover and Quantitative Imaging (QI^{TM})) are presented.
In contact mode, a setpoint repulsive force (or vertical deflection of the cantilever) is fixed and the probe is moved from point to point. If the force changes for the new probe position, the z position is adjusted, so the setpoint force is again reached. In this mode, the probe is permanently in contact with the surface. This means that, if there is a change in the height of the sample, the exerted force will be larger than the setpoint, before the z position is adjusted. Also, as the probe is dragged on the surface, it experiences a lateral deflection of the cantilever, exerting a torque. Both cases are detrimental to the probe apex condition, and the integrity of the sample. As the probe scratches the surface, the probe apex can become blunt, reducing the resolution and volatile parts of the sample can be removed by the probe. Overall, this mode causes strong wear effects for both probe and sample.
This wear can be reduced by measuring in AC mode, in which the probe is driven to oscillate at a frequency slightly below the resonance frequency. Approaching the surface with the probe, the atomic interactions shift the resonance frequency of the oscillation, which can be detected by a change in e.g. the amplitude of the oscillation. The situation is depicted in Fig. 2.5. The amplitude of the shifted oscillation is kept constant by the feedback system during scanning.
In this mode, the probe only touches the sample for a brief moment at one extremal point of the oscillation, giving this mode its name intermittent contact or tapping mode. As the probe is not permanently in contact with the sample, lateral deflections of the cantilever and overshoots in force are reduced, in turn reducing wear effects on the probe and the sample. The problem of measuring in AC mode is, that the exact force which is applied when touching the surface is not well defined.
A variation of AC mode is the so-called hover mode used, in the scope of this thesis, to acquire magnetic force microscopy measurements (see Subsec. 2.2.4). Every line of an acquired image is scanned in both directions, exploited in hover mode, to extract additional information. For every line, one direction measures the topography of the sample in AC mode. At the turning point, the probe is lifted by a specified, constant z offset and follows the z profile measured before, on the same line. This way, changes in amplitude and phase due to atomic forces are minimized in the return profile and forces caused by other effects (such as magnetic forces) can be seen as changes in the phase of the oscillation, while ensuring that the tip will not crash into the surface.

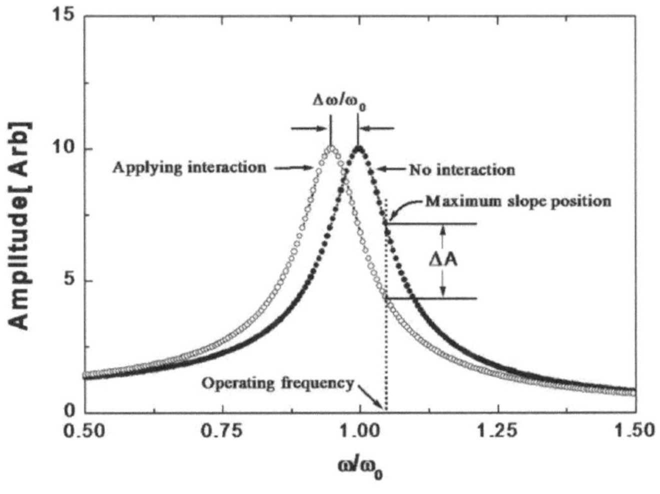

Figure 2.5: Operation principle of AC mode. Atomic interactions between probe and sample shift the resonance frequency of the cantilever oscillation by $\delta\omega$, resulting in a change in amplitude of ΔA (reprinted with permission from [17] ©2022 Park Systems).

Analogously to STM, spectroscopic measurements are possible in AFM. The most intuitive spectroscopic measurement is of the force upon approaching the surface, an exemplary measurement of which is shown in Fig. 2.6. In these measurements, the probe is positioned at a fixed lateral position and is moved closer to the sample, measuring the force during the approach and the retraction of the probe, up to a previously set maximum force.

When the probe is far away from the sample (A), the deflection of (or force on) the cantilever is constant, so a linear fit to the deflection in this region serves as the subtracted baseline for the force–distance curve. Once the probe is close enough to the sample, the attractive forces between probe and sample pull the probe into contact (called 'snap in', B), bending the cantilever towards the surface, enhanced by the water layer that forms in ambient conditions. The probe is approached further, linearly increasing the vertical deflection up to the set maximum force (C). A linear fit to this region yields a slope that is representative of the stiffness of the sample and the cantilever. The extraction of the sample stiffness from a force–distance curve is possible, using an appropriate model, such as the Hertz model [18–20]. As the stiffness of the cantilever is the same at every point, mapping the slope is representative of the changes of the sample stiffness.

Retracting the probe, the same behavior is observed, as long as the probe is still in contact. The adhesion forces between probe and sample lead to a larger bend of the cantilever towards the surface (D), compared to the snap in, until finally, the

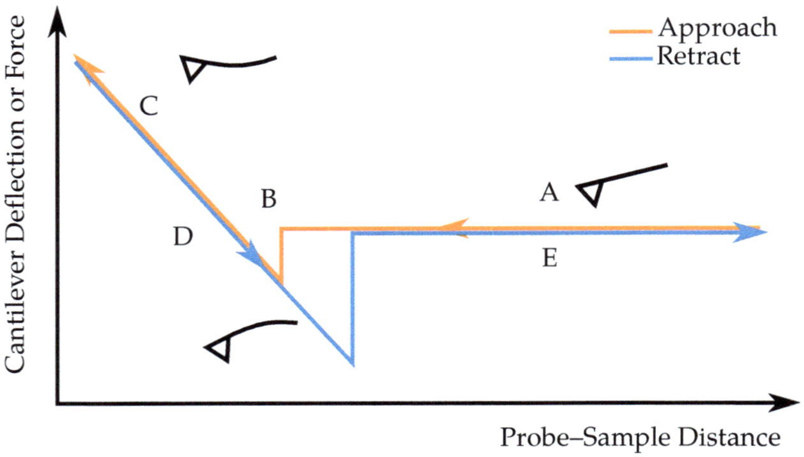

Figure 2.6: Typical force–distance curve. When the probe is far away from the sample it is flat, defining the baseline. At a certain distance the probe 'snaps in'. After it crosses the baseline force is applied up to a setpoint, the slope being a measure of the stiffness of the sample (and cantilever). Upon retraction, the probe sticks to the surface until it 'snaps out', which is a measure for adhesion, and returns to the baseline.

contact breaks (called 'snap out', after D). The extent to which the force overshoots in the retract curve is a measure of the probe–sample adhesion. After the snap out, the deflection returns to the baseline (E). In addition to measuring the force, other quantities can be extracted simultaneously, such as the current in conductive AFM. Finally, Quantitative Imaging (QI) mode is a combination of scanning and force spectroscopy, in which the surface is studied point by point, but at each (x,y) position, a force–distance curve is taken. All the curves are measured up to a fixed maximum force and, analogous to contact mode, mapping the z position at that maximum force represents the topography. There are several advantages of QI over the other presented modes. Firstly, the force applied at every point can be precisely controlled, removing this uncertainty compared to AC mode. Secondly, the probe is still retracted far enough from the surface, so that wear effects are minimal. Finally, other quantities such as adhesion, stiffness (slope), and, specifically in conductive AFM, extremal currents can be mapped easily. The disadvantage of QI mode is the increased measurement time per scan, as the timings of force–distance curves have to be appropriate and take longer than adjusting z with the feedback loop. However, the advantages make QI the favored operational mode in the scope of this thesis.
It is especially useful in conductive AFM, where great care has to be taken that the probe coating remains intact and it is useful to map the extremal current, as discussed in Subsec. 2.2.3.

2.2.3 Conductive AFM

Conductive AFM (CAFM), also called conducting probe AFM or current-sensing AFM, is a variation of AFM, where both the sample and the probe are conductive. Other methods using this kind of setup, such as tunneling AFM (TUNA), scanning capacitance microscopy (SCM) and Kelvin probe force microscopy (KPFM) are not discussed in more detail, here.
In CAFM, probe and sample are in direct contact and an applied bias voltage leads to a current, which obeys

$$I = JA_{\text{eff}}\,, \tag{2.12}$$

in which J is the current density and A_{eff} is the effective contact area. This is a virtual area including all parts of the surface, which are electrically connected to the probe, varying heavily, depending on the probe radius, the sample conductivity and the load force. For highly conductive samples, such as metals, the effective contact area can spread over large parts of the sample, due to its high lateral conductivity. For samples with lower conductivity, this area becomes smaller, increasing the resolution of CAFM.
A common problem in CAFM is the conductivity of the sample and probe. For the sample, conductivity can be reduced strongly due to contamination of the surface, occurring comparably quickly, especially in ambient conditions. In case of the probe, contamination also reduces the conductivity, but the more prominent issue is, that probes are usually made from non-conducting material (often Si) coated with a metal (Pt, PtIr, Au, etc.). This coating has to be thin, to keep the probe radius small for better resolution, meaning that the coatings are very volatile and easily removed or worn off and have to be handled carefully (see Subsec. 3.3).
An advantage of CAFM over STM is that the topographic and the electronic properties can be studied separately, as the topography information comes from the force. This makes the interpretation of CAFM data more straightforward. However, the resolution of AFM is lower than for STM, because the distance dependence of the current for the tunneling effect from the terminating atom of the STM probe apex is very strong, in contrast to the atomic forces between an AFM probe of comparably large radius with the surface.

2.2.4 Magnetic Force Microscopy

As mentioned in Subsec. 2.2.2, it is also possible to extract information on magnetic properties of samples with AFM, which is termed magnetic force microscopy (MFM). For this technique to work, a magnetic probe is necessary. The most commonly used probes are Co– or CoCr–coated Si probes. As magnetic forces between the probe and the sample can also induce bending of the cantilever or a shift in the resonance in AC mode, they can be detected in an AFM setup. The z component of the force F_z on a magnetic probe in a magnetic stray field can be described by [21]

$$F_z = -\mu_0 \int M_{z,\text{probe}} \frac{\partial H_{z,\text{sample}}}{\partial z} dV \,, \tag{2.13}$$

where μ_0 is the vacuum permeability, $M_{z,\text{probe}}$ is the probe magnetization, $H_{z,\text{sample}}$ is the sample magnetic stray field and V is the volume of the magnetic material of the probe. As mentioned in Subsec. 2.2.2, the MFM signal is a difference in phase due to a shift of the resonance frequency of the cantilever. The phase shift $\Delta\phi$ is related to the gradient of the magnetic force F_z' [22] *via*

$$\Delta\phi = -\frac{Q}{k_\text{cant}} F_z' \,. \tag{2.14}$$

One problem is that magnetic forces are small, compared to the atomic forces, when the tip is in contact. Moreover, since both exert a force on the probe, they cannot be measured separately in a scan. These problems are mostly solved by measuring in hover mode, as explained in Subsec. 2.2.2. If the height offset for the return profile is chosen too small, remainders of the atomic forces can still be visible ($F_\text{VdW} \propto z^{-7}$ for short-range interactions [23]), if it is chosen too large, the signal caused by the magnetic field is not measurable ($F_\text{mag} \propto z^{-4}$ [24]). Typical height offsets for a observable magnetic contrast in the return profile are $50 - 100$ nm, as tip–sample forces due to VdW interaction also have long-range components ($F_\text{VdW} \propto z^{-2}$), according to [25].
Having described the principles of the measurement techniques used, the following sections focus on a brief history and potential drives of molecular motors, as studied with STM in this thesis.

2.3 Molecular Motors

The first studied systems involving molecular motors are biomolecules, such as Kinesin [26] and Myosin [27], needed in the process of converting chemical energy in the form of adenosine triphosphate (ATP) to motion. Together with other proteins, they are responsible for intracellular transport along the structures of the cytoskeleton of living cells. Biomolecules performing rotary motion, such as bacterial flagella, act as propellers to move in liquids [28]. They can transform a proton gradient into rotational motion.

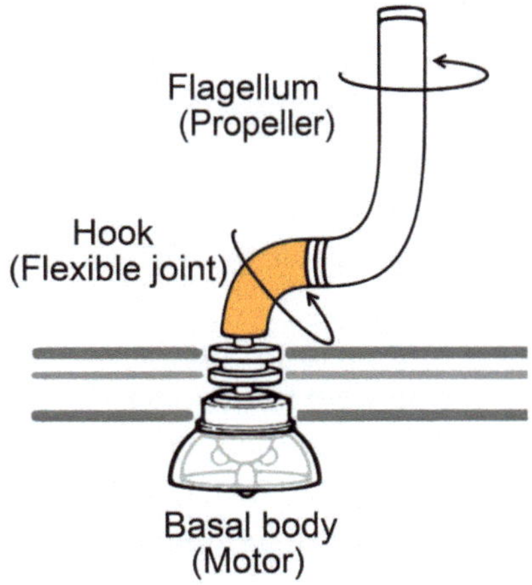

Figure 2.7: Sketch of a flagellum which is able to perform rotational motion driven by a proton gradient, so that bacteria can move in liquids (taken from [29]; Credit: Dr. Satoshi Shibata, OIST under CC BY 2.0 license[2]).

Inspired by the success of the investigation of these molecules and by the ever decreasing size of artificial nanostructures, Feringa *et al.* [2] and Kelly *et al.* [30] realized the first artificial molecular motors. They are much smaller than the bulky proteins and are driven by optical and chemical means. In the chemically driven molecule, formation and breaking of chemical bonds between parts of the molecule causes rotation. In the light-driven molecular motor, different conformations are cycled by irradiation and thermal activation.
Another approach was to drive a molecular motor electrically by Sykes *et al.* who reported asymmetric rotation rates for different rotation directions, observed in isomers of butyl methyl sulfides. They attributed the asymmetry to an intrisic chirality of the STM tips used in their study [31]. In other reports, very small molecules rotated on asymmetric substrates [32] and self-decoupled molecular motors with a 'ball bearing' were being rotated by tip manipulation [33].
In this work, a slightly different approach is chosen. As described in [33], the rotor of the molecule is supposed to be decoupled by means of a platform which both fixes the molecule on the surface and decreases the interaction between the rotor and the substrate. However, the tip is to be held at a fixed position during the rotation process. Injecting a current from the STM tip should then lead to directed motion of the rotor.

[2]`https://creativecommons.org/licenses/by/2.0/legalcode`

2.4 Proposed Drives for Directional Rotation

For the realization of an electronically driven molecular motor, it is important to first consider how the injected current can be transformed in such a way that the rotor performs directed rotation. In addition, the rotational motion has to happen in a step-wise fashion for it to be observable in STM, which is not straightforward in nanoscale systems.

The restriction to step-wise rotational movements is due to the experimental limitations of STM concerning measurable frequencies. Towards low frequencies, the time resolution limit is set by the drift. Small temperature variations cause minute expansion or shrinkage of the parts of the STM, which leads to relative shifts between tip and sample. Therefore, to ensure that long measurements are taken at the same relative tip–molecule position and to avoid significant drift of the tunneling current during measurements, frequencies of less that $\sim 10^{-2}$ Hz are not feasible. For such low frequencies, obtaining a large enough statistic sample would take so long, that the drift would be too large to allow for comparison of the beginning and end of a measurement.

Towards high frequencies, STM measurements are limited by the bandwidth of the transimpedance amplifier (see Sec. 5.3), which is $\sim 10^3$ Hz. Any fluctuations or current changes at higher frequencies are attenuated strongly and cannot be observed.

Comparing these limitations with typical frequencies of fluctuations within molecules, such as vibrational modes, which are on the order of $\sim 10^{12}$ Hz, it becomes clear, that they cannot be observed directly. A possibility to follow the rotation in STM is to observe switching between metastable states assumed during the rotation process. Therefore, the molecular design has to be such, that metastable states can be assumed and have lifetimes corresponding to the measurable frequency range of STM. These lifetimes can be estimated using a simplified transition state theory.

In addition to the limitation in frequency to allow for observation of a rotation, a drive is needed to induce a directed, not a random rotation, for which two possibilities are presented in this section. Also for the drive, the STM sets the boundaries of parameters to excite directed rotation. For example, if the drive were the Lorentz force, rough estimates lead to very extreme parameters (approximately 10 nA and 6 T) to be able to get a rotation which significantly deviates from random motion. Therefore, the drive chosen here is the chiral-induced spin selectivity (CISS) effect. The basics of this effect and how it can be used to drive a molecular motor are outlined in Subsec. 2.4.1, followed by an alternative approach to driving a molecular motor, namely the on-off-ratchet in Subsec. 2.4.2.

2.4.1 Chiral-Induced Spin Selectivity

Already as early as 1908 C. W. Richardson theoretically outlined an effect which connects the magnetization of a material with an angular momentum [34]. This was proven in 1915 by A. Einstein and W. J. de Haas. They studied the effect of a magnetic field on the rotation of a magnetizable cylinder [35]. In their experiment, a cylinder made of nickel or iron is placed in the center of an electromagnetic coil. Upon running a current through the coil, the spins of the atoms in the cylinder align with the magnetic field leading to a net spin angular momentum. However, the total angular momentum of zero of the cylinder cannot change due to the conservation of angular momentum. Therefore, the cylinder performs rotational motion which can be detected.

In 2007, T. Ono and H. Kohno [36] proposed to drive nano-machines exploiting a similar effect. In their proposed motor, a spin-polarized current is to run through the nano-machine. It relaxes in the machine and loses its spin-polarization. Again, due to the conservation of angular momentum, the machine gains the equivalent of the change in spin-polarization in angular momentum. This is supposed to lead to a rotation of the nano-machine.

In the scope of this thesis, the inverse effect is proposed. A non-spin-polarized current is to be injected into the nano-machine (a molecule in this case). The current is polarized within the molecule and the opposite of the equivalent in angular momentum is transferred to the molecule, driving a directed rotation.

The way to polarize the spin of the injected current is chosen to be the chiral-induced spin selectivity (CISS) effect, due to the comparably compact molecular design necessary to realize it in a molecule. In CISS, the spin of a current is selectively transferred through a molecule, depending on its chirality [37]. In transport, this means, that if a non-spin-polarized current is injected into a chiral molecule, the current exiting the molecule has a certain degree of spin-polarization.

The CISS effect can be explained qualitatively in an intuitive way. Assuming a helical molecule, which is necessarily also chiral, an electron passing through the molecule interacts with its charge distribution. In the resting frame of the electron, the helical charge distribution of the molecule moves around the electron, resulting in the shape of a ring current. It therefore causes a magnetic field, which interacts with the spin of the electron, defining a preferred spin orientation, effectively causing a spin-polarization. In more sophisticated descriptions of the effect, often a linear chain of atoms in the shape of a helix is assumed, frequently including a tight-binding approach to simulate transport through the molecule [38, 39].

Although spin-dependent transport through helical molecules is observed in many experiments [40–43], its origin remains an open question. As reviewed recently [44], various routes of exploring the CISS effect have been followed experimentally

and theoretically. While most theories do yield a qualitative spin-polarization or spin-filtering effect, they do not explain CISS quantitatively, predicting the effect to be of much smaller extent than observed in experiments. In most cases, this is due to the description needing spin-orbit-coupling (SOC), which is comparably weak for light elements, such as carbon, nitrogen and oxygen, the main building blocks of organic, helical molecules.
In a recent approach by F. Evers and J. M. van Ruitenbeek [45], the CISS effect is described by a cylindric free-electron gas, in which interacting centers for only SOC are placed in a helical arrangement, to model the atomic nuclei of a chiral molecule. Through scattering at these centers, an orbital momentum is transferred to the centers, in turn causing a spin-polarization of the electrons. The advantage of this approach is the disproportionality between the mass of the electron and the nuclei, effectively replacing the electron mass by the moment of inertia of the molecule in the calculation of rotational energy, which allows for a strong spin-polarizing effect on the electrons and only a small rotational movement of the nuclei.
Aside from the open questions regarding the cause of the CISS effect, it is also unclear, whether it can be exploited in a molecule, as proposed in [36], to transfer an orbital momentum and a torque, to drive a directed rotation.

2.4.2 On-Off Ratchet

Another way to drive directed motion is based on the on-off ratchet, or flashing Brownian ratchet [46–48] and needs an asymmetric potential with local minima, as sketched in Fig. 2.8. Looking at the rotor as a particle, the probability density to find the particle in one of the minima can be described as a wave packet therein. As long as the barriers in both directions are higher than the particle energy, this wave packet remains in its potential well.
For directed motion, the potential barriers have to be lowered and raised repeatedly. In the situation with lowered barriers, the particle has enough energy to cross the barriers. The wave packet is not confined to the potential well anymore and disperses symmetrically.
By raising the potential barriers again, parts of the wave packet will be located in the adjacent potential wells. Since the potential is asymmetric, more of the wave packet will be located the well in one direction, compared to the other. Therefore, the probability density to find the particle in the potential well on one side of the initial well is larger than for the other side. If the potential barriers are repeatedly lowered and raised, this should result in a net motion of the particle in one direction.

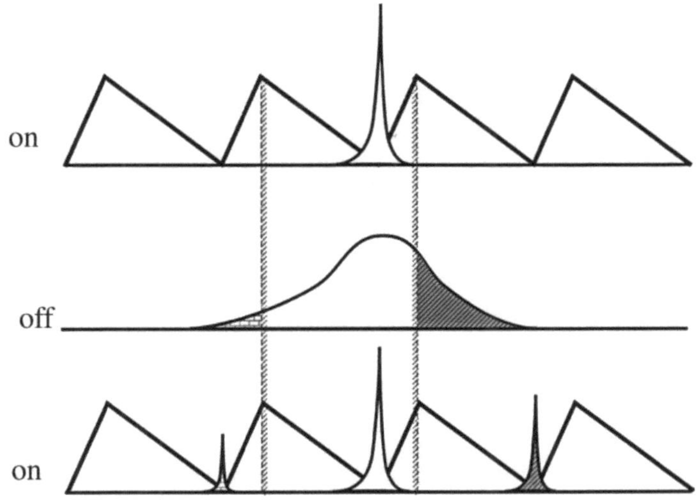

Figure 2.8: Principle of the on-off ratchet. Turning an asymmetric potential on and off periodically can lead to directed motion through the symmetric dispersion of a wave packet (reprinted with permission from Taylor & Francis Ltd. from [46]).

2.5 Principles of Evaluation Methods

Before going into more detail about the experimental setup, it is important to outline the basics for some of the methods used in the evaluation of data. These are, the binomial test and mean square displacement (MSD). As it will become evident, the binomial test can help to judge, whether the detected motion is directed or random. The MSD is a useful tool to interconnect the measured data to actual physical properties and to verify other methods of determining e.g. the velocity of rotation.

2.5.1 Binomial Test

The binomial test is a commonly used approach to decide if the distribution of two possible outcomes of an experiment is of statistical significance. It tests, based on a sample from the distribution, whether an assumed null hypothesis can be rejected. Note, that this does not mean, that the exact opposite statement is true.
The first step in the binomial test is to formulate the null hypothesis. This is the statement which is supposed to be rejected with a certain significance level. It is defined by the probability of success p_+ of one of the two possible outcomes in one instance of the experiment. Once this probability is defined, the binomial test provides the means to verify, whether the sample data is sufficient to reject the null hypothesis.

Every instance of the experiment has two outcomes, success or failure, which occur with probabilities p_+ and $1 - p_+$, respectively. Repeating the experiment n times yields n_+ successes. The binomial distribution gives the probability of obtaining n_+ successes, which can be calculated by

$$\Pr(n_+) = \binom{n}{n_+} p_+^{n_+} (1 - p_+)^{n - n_+} . \tag{2.15}$$

In a one-tailed test the p-value is computed by summing up the probabilities Pr of more extreme results than measured:

$$p = \sum_{i=0}^{n_+} \Pr(i) = \sum_{i=0}^{n_+} \binom{n}{i} p_+^i (1 - p_+)^{n-i} . \tag{2.16}$$

This case would be a left-sided test. The right-sided test can be performed analogously with a summation from n_+ to n.
This calculation can be used in two different ways. One is to fix the significance level $p = \alpha_0$ to a desired threshold and compare the experimentally determined p-values to evaluate whether the data rejects the null hypothesis with this significance level. Commonly used values are 0.01 and 0.05. In fields like particle physics, the criteria are much stricter, setting the significance level to 5σ, which corresponds to $\alpha = 3 \times 10^{-7}$.
The other way is to directly use the p-value, which provides a measure of how extreme the measured value is compared to the expected values assuming the null hypothesis. Therefore, a small p-value means that the observed number of successes would be observed very unlikely if the null hypothesis were true. To convert the p-value into a measure of how probable it is to reject the null hypothesis, the so-called 'rejection probability' p_{reject} is introduced. It can be calculated as

$$p_{\text{reject}} = 1 - p \tag{2.17}$$

and is usually given in percent. As becomes apparent in Chap. 7, this proves a useful tool to judge, whether directed rotation is observed, rather than random motion. Similarly, it will be useful for the characterization of the observed rotation to calculate the mean square displacement, which will be introduced in the Sec. 2.5.2.

2.5.2 Mean Square Displacement

The mean square displacement (MSD) is a quantity often used in the description of random motion, such as Brownian motion of particles. It finds application in the description of biological processes, fluid mechanics and diffusion of atoms on surfaces [49–53]. The MSD can be used to extract information on the nature of motion,

e.g. whether it is random or directed. Moreover, parameters, such as the diffusion constant, velocity and, if applicable, anomalous diffusion constants can be extracted from mean square displacement plots [54, 55]
In its original form, it is described as an ensemble average of the squared displacements with respect to a reference point of particle trajectories $r_n(t)$ at different lag times τ

$$\mathrm{MSD}(\tau) = \langle r^2 \rangle(\tau) = \lim_{N\to\infty} \frac{1}{N} \sum_{n=0}^{N} \left[r_n(\tau) - r_n(0) \right]^2 , \tag{2.18}$$

where lag times τ are times after the starting point $t = 0$. In this description, the MSD is a measure of how far the particle has travelled during lag time τ.
Assuming ergodicity, the ensemble average and the time average over a single trajectory are the same. Since the measurements performed in this work are time-dependent and not ones of many particles at the same time. As different measurements cannot be compared easily, a calculation of the MSD via a time average is more convenient. Therefore, Eq. (2.18) can be rewritten as

$$\mathrm{MSD}(\tau) = \lim_{T\to\infty} \frac{1}{T} \int_0^T \left[r(t+\tau) - r(t) \right]^2 dt . \tag{2.19}$$

Calculating this quantity for all lag times τ allows for an observation of the evolution of the MSD. Plotting MSD versus τ, the diffusion constant D, the velocity v_{MSD} of a directed motion and the anomalous diffusion constant Γ_α can be extracted, if they are present. This is possible because the MSD can also be expressed in other terms, according to [55]

$$\mathrm{MSD}(\tau) = 2n_{\text{dim}} D\tau + (v_{\text{MSD}}\tau)^2 + 2n_{\text{dim}} \Gamma_\alpha \tau^\alpha , \tag{2.20}$$

where n_{dim} is the dimensionality of the movement of the particle. If the particle moves purely diffusively, the MSD should follow a linear trend (red line in Fig. 2.9). While for a directed motion (blue line), the MSD shows a superimposed quadratic behavior on top of the linear behavior due to diffusion. Anomalous diffusion describes a behavior which is neither diffusive, nor directed, characterized by the exponent of the lag time dependence α. Depending on the origin of this diffusion, it can have values of $0 < \alpha < 1$ (subdiffusion) or of $1 < \alpha < 2$ (superdiffusion).
With this connection, plotting the MSD provides a means to extract physical quantities from measurements, describing diffusive and/or driven behavior. This can help to discriminate between directed and random motion and yields values for comparison with other molecular machines. Exemplary plots of the MSD and the respective behavior they describe are shown in Fig. 2.9.

[3] https://creativecommons.org/licenses/by-sa/3.0/legalcode

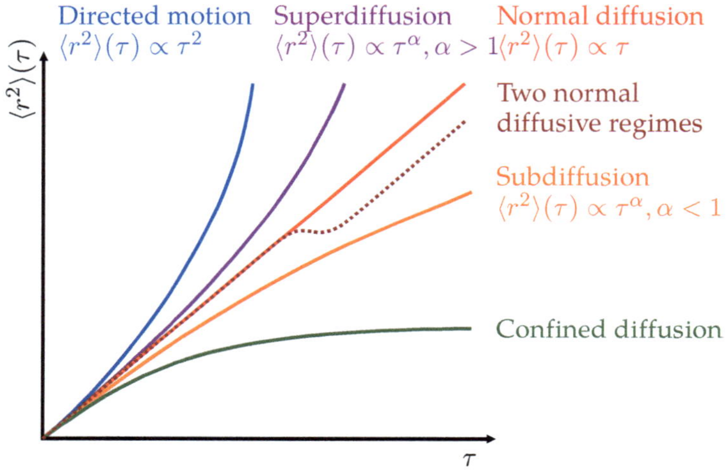

Figure 2.9: Sketch of different diffusive cases in MSD vs τ plots. The different behavior of the MSD depending on the lag time gives insight into the underlying processes of diffusion of the particle (adapted from [56] under CC BY-SA 3.0 license[3]).

3 Experimental CAFM setup

Section 2.2 presents fundamentals of conductive atomic force microscopy (CAFM), which is used to study self-assembled monolayers of chiral molecules, to get insight into the CISS effect, which is a proposed drive for the directed rotation of the molecular motor presented in Chap. 7.
Here, the experimental setup, that implements the discussed fundamentals and was used to perform the measurements shown in Chap. 4, is presented, comprising descriptions of the different elements in the AFM setup (Sec. 3.1), the specific parts needed for CAFM (Sec. 3.2) and the types of probes that were used for CAFM studies (Sec. 3.3). This setup was used during a stay at Leiden University, in the group of Prof. Jan M. van Ruitenbeek and is part of the Scanning Probe Microscopy Facility at the Leiden Institute of Physics (Leids Instituut voor Onderzoek in de Natuurkunde, LION).

3.1 AFM Setup

The setup used to perform CAFM studies is a commercially available *JPK NanoWizard®* 3 system[1]. It measures the cantilever deflection by the deflection of a laser beam pointed at the backside of the cantilever (compare Subsec. 2.2.1) and is operated in ambient conditions (atmospheric pressure and room temperature).
Figure 3.1 shows an overview of the complete setup. The AFM is housed in a soundproof chamber, to reduce vibrational noise from the environment, improved by covering the interior with sound-absorbing foam. It also serves as a Faraday cage to reduce in-coupling of noise currents. Vibrations are further reduced by letting the AFM stand on an active vibration isolation system by *Accurion* [2].
Placed on the active vibration isolation is a stand to which a microscope is fixed. The microscope points from the top on the backside of the cantilever and has a variable magnification of up to $\times 7$. It is also used to find the sample and roughly

[1] `https://www.bruker.com/en/products-and-solutions/microscopes/materials-afm.html`
[2] `https://accurion.com/active-vibration-isolation/products/i4-series`

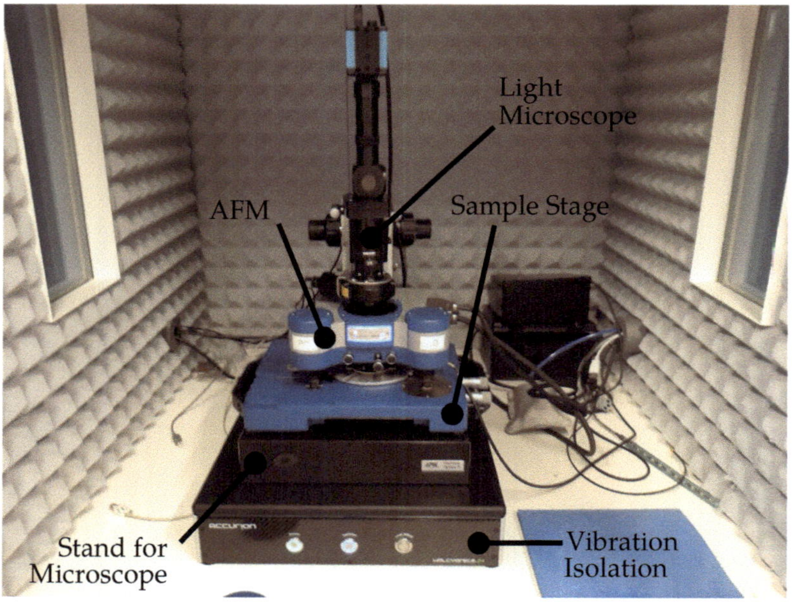

Figure 3.1: View of the AFM setup. Inside the soundproof chamber, the AFM stands on an active vibration isolation system. The AFM (blue) is mounted on a stand, connected to an optical microscope (black, above AFM), which can image the cantilever backside and the sample below for rough positioning of the sample.

(a) Top view of the sample stage with sample holder (dark blue). It can be moved over large distances in the (x,y) plane to find suitable spots to measure.

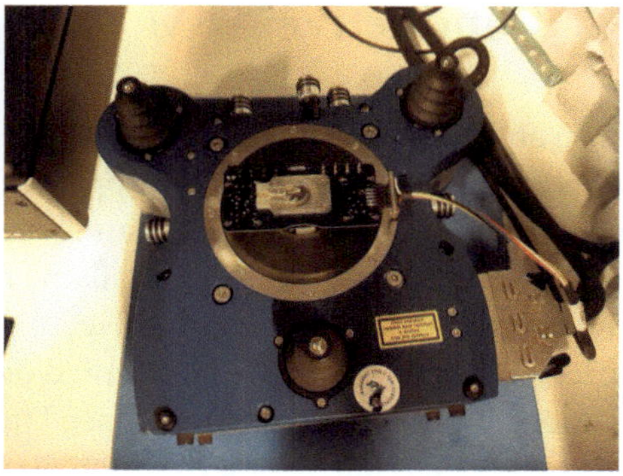

(b) Bottom view of the AFM, showing the CAFM tip holder and the three feet for coarse z motion.

Figure 3.2: Detailed image of sample stage and bottom side of the AFM.

find a suitable area for measurements thereon. Moreover, it is used to correctly adjust the position of the laser spot to the backside of the cantilever. It is equipped with two different light sources, a LED ring light and a centered sodium-vapor lamp for illumination. Finally, a CCD camera transfers the obtained image to acomputer.
The sample stage can be moved with stepper motors in the (x,y) plane to find a good probe position for measurements. It has suited indentation for fixing the sample holder (dark blue) by two clamps, shown in Fig. 3.2a. The AFM stands on the sample stage with three feet, which can move the complete AFM up and down using stepper motors, for approaching the probe to the sample surface. They are shown in Fig. 3.2b, depicting the bottom side of the AFM. Here, also the CAFM tip holder (described in Sec. 3.2) can be seen, connected (colored cable) to the control unit.
The probe is moved by piezo actuators in all three directions. In this system, the maximum movement range is 30 µm laterally and 6.5 µm in z. The lateral resolution is mainly defined by the probe radius and meaningful scans of sizes smaller than 100 nm can be challenging. However, the AFM can be precise down to the sub-nanometer scale in z. As the measurements are performed in ambient conditions, drift is another limiting factor, especially laterally.
Due to fluctuations in room temperature, materials expand or shrink, causing movement of all parts, including the cantilever, probe and sample. As these fluctuations are relatively large at room temperature (i.e. compared to low temperatures, such as for the STM setup, Sec. 5.3) and because SPM imaging is a slow process, they can cause distortions and shifts between successive images.
With the basic AFM setup described, Sec. 3.2 presents the necessary equipment to perform conductive AFM measurements.

3.2 Conductive AFM

To be able to perform conductive AFM (CAFM) measurements, two more specific parts are needed to allow for applying a bias voltage to sample and probe and measure the current flowing, namely a specific probe holder and sample holder. These parts are commercially available from *JPK*.
The probe holder is depicted in Fig. 3.3a. It is connected to the control unit *via* the port on the left. The probe is mounted on the central glass part by a clamp, which is connected to a square gold pad (see Fig. 3.3a) and held in place by being a spring inside the glass part. For a conductive probe, it also provides the electrical contact to the preamplifier, which is directly on the back side of the probe holder.
The preamplifier is mounted directly on the probe holder to reduce cable length, in which the small measured current is carried, because here, the current measurement is most susceptible to noise. Even small noise currents coupling into the system at this point are non-negligible, as they are still comparable to the measured currents, which are in the range of a few pA up to 12 nA. These currents are converted to measurable voltages by the preamplifier. The saturation current of 12 nA is defined by the the preamplifier's feedback resistor, which also defines the amplification. Here, it is $R_{\text{feedback}} = 1\,\text{G}\Omega$, corresponding to an amplification of $10^9\,\frac{\text{V}}{\text{A}}$.
The CAFM sample holder is shown in Fig. 3.3b. The sample is placed in the center of the non-magnetic metallic plate. A clamp presses onto the conducting surface, establishing electrical contact, itself being connected to the gold pad. A thin, flexible wire is soldered onto the gold pad and to a connector that fits onto the pins of the CAFM probe holder (bottom left in Fig. 3.3a). The connection allows to apply a bias voltage and measure a current through the probe and sample. The voltage is applied to the sample, while the probe is kept on ground, to prevent detrimental effects on the tip coatings. That is why the currents presented in Chap. 4, according to the definition of the direction of current in the *JPK* software, are negative.
All current maps presented in Chap. 4 are obtained using QI mode. To get easy access to current maps, the minimal current is directly extracted from the current–distance curves, measured simultaneously to the force–distance curves during QI imaging. However, the single curves can still be viewed for every pixel. For QI imaging, it is important to adjust the timing correctly, as scanning too fast can cause artifacts in the force–distance curves. A main issue is the return point from approaching to retracting. If the spectroscopy is done too fast, the piezo actuator is still expanding, when the data for the retracting curve is already being taken, leading to an overshoot in force in the approach curve. A typical time for a single force–distance curve in a QI image is $t_{\text{force–distance}} \approx 30\,\text{ms}$. Therefore, QI measurements are comparably long ($\sim 9\,\text{min}$ for one image), increasing effects of e.g. drift.

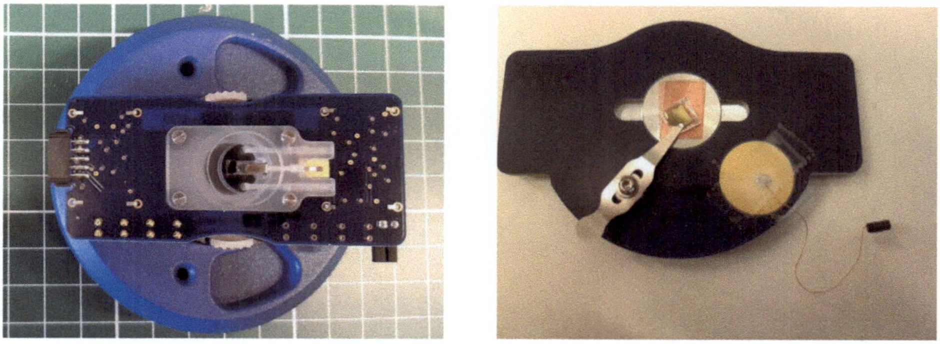

(a) CAFM probe holder. The probe is held in place by a metal clamp (center), also making contact to the preamplifier. The cantilever is positioned on the clear glass part, which lets the laser beam pass to the cantilever.

(b) CAFM sample holder. The clamp connects the sample electrically to the gold pad. The cable connects to the gold pins (bottom left, Fig. 3.3a) of the CAFM sample holder to apply a voltage. The complete part is non-magnetic.

Figure 3.3: Detailed image of CAFM probe and sample holder.

3.3 CAFM Probes

The fundamental requirement for an AFM probe to be usable in CAFM is electrical conductivity. For that purpose, two different probes are used in the measurements presented in Chap. 4, CoCr-coated Si probes and Pt wire probes.
The Pt wire probes are commercially available RMN-25PT300B probes from Rocky Mountain Nanotechnology [3] (RMN). A schematic of the chip, on which the wire is mounted, is shown in Fig. 3.4a. The wire is glued to a rectangular alumina chip and electrically connected to a pad of Au. The clamp of the CAFM probe holder is pressed thereon, to connect it to the preamplifier.
The cantilever is specified with $L = 300\ \mu\mathrm{m}, b = 100\ \mu\mathrm{m}, k_{\mathrm{cant}} = 18\ \frac{\mathrm{N}}{\mathrm{m}}$ and $f_{\mathrm{resonance}} = 20\ \mathrm{kHz}$, however, variations can be large, as every cantilever is manufactured uniquely. In fact, thermal calibration of the actual spring constant is difficult, because the fitting procedure of the *JPK* software often could not fit the resonance properly, as the peak was cut off and flat. Even if the fit could be performed, the results deviated strongly from the specifications, usually being at $f_{\mathrm{resonance}} \approx 27\ \mathrm{kHz}$ and $k_{\mathrm{cant}} \approx 5\ \frac{\mathrm{N}}{\mathrm{m}}$. This implies that the forces applied with RMN probes are not very well defined and are most likely higher than the ones displayed by the software.
The expected advantage of the solid Pt wire probes is that they are always conductive, as there is no volatile coating which can be worn off from an otherwise insulating probe. Also, Pt is a rather inert material, so the formation of insulating oxide layers

[3] `https://rmnano.com/tech-data`

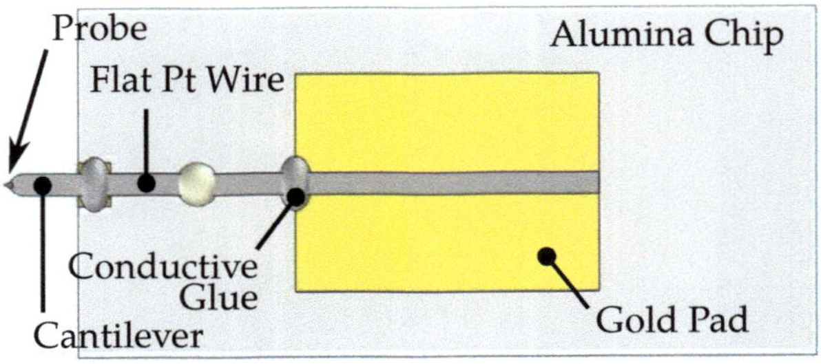

(a) Schematic of the chips for RMN Pt wire probes. The gold pad is connected to the preamplifier by the CAFM probe holder clamp.

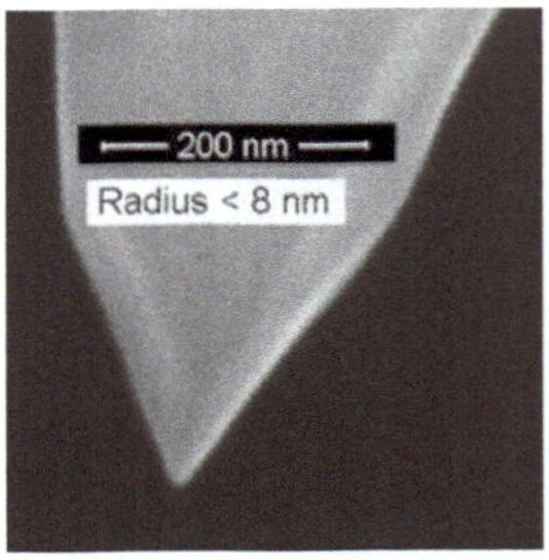

(b) RMN Pt wire probe apex. Its apex radius is specified to be $<$ 20 nm.

Figure 3.4: RMN solid Pt wire chip and probe characteristics (reprinted with permission ©2022 Bruker and 2021 RMN [57]).

should be minimal. However, the used probes were relatively old (~ 4 y) which led to the observation, that they were still insulating in some cases. However, preparing them on a bare Au surface allowed for current sensing, with the downside of a certain volatility towards long measurements, in which the tip sometimes changed and became insulating.
The CoCr-coated Si probes are MESP-V2 probes bought from *Bruker*[4] in different variations (normal, low magnetic moment or LM and high magnetic moment or HM). The CoCr coating is magnetic, making the probes useful for both CAFM and MFM measurements. The different types have various thicknesses of CoCr coating, being 17, 27 and 72 nm for MESP-LM-V2 (LM), MESP-V2 (normal) and MESP-HM-V2 (HM), respectively. This gives the probes different magnetic moments and coercive magnetic fields, increasing from LM to normal to HM. Exemplary SEM images of the chip and cantilever and the probe are shown in Fig. 3.5.
These probes are chosen for comparability reasons, because self-assembled monolayers of chiral molecules were studied in other publications, with these probes [41, 42]. Also, the magnetic coating can potentially be used to inject a spin-polarized current to study the effect of chiral molecules on it.
The problem of this type of probe is the coating. For one, it is rather thin, making it volatile and easy to wear off. Figure 3.6 shows images of a pristine and two damaged probe apexes. With all of the probes shown therein, no current could be measured. Upon trying to measure a current, rather large forces were applied, which led to the destruction of the apex. Also, probe coatings are usually not perfectly uniform and do not exhibit the same conductance at every point of the CoCr–Si contact.

[4] `https://www.brukerafmprobes.com/p-3948-mesp-v2.aspx`

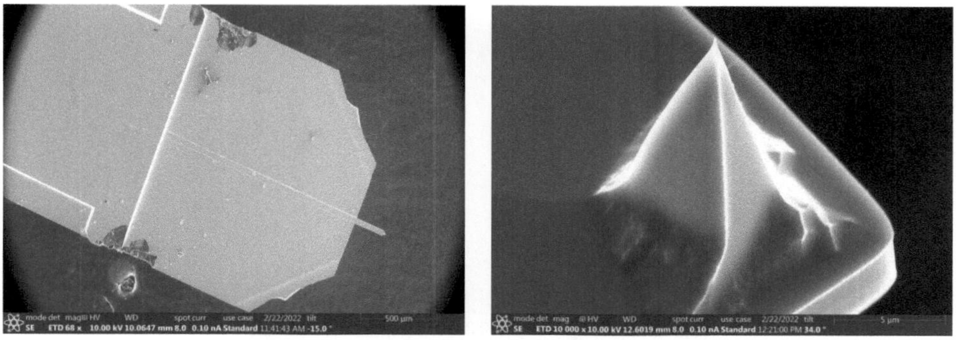

(a) MESP-V2 chip. The cantilever with the probe points towards the right.

(b) MESP-V2 probe. It has a pyramidal shape with a very sharp apex.

Figure 3.5: SEM images of a MESP-V2 chip and probe.

Therefore, if a voltage is applied between the sample and the probe, the current can run through very small areas at the CoCr–Si interface and partially melt the coating. Furthermore, inappropriate handling of the probes can lead to static discharges, removing the coating partially or completely. Figure 3.6 shows the volatility of the coating.

Moreover, conductivity of the probes can deteriorate quickly. As Co is a rather reactive material, an oxide layer can easily form on the surface of the tip coating, reducing the conductivity. In fact, for the probes used here, it was only possible to obtain reasonable currents with fresh probes or at very high forces. After being exposed to air, even though stored in rough vacuum for most of the time, the measurable currents were reduced strongly, rendering it impossible to obtain current maps. A different cause for the reduction of conductivity, especially when working on molecules, is contamination with surface adsorbates that are picked up by the probe apex. If they are attached to large parts of the probe apex or even form multilayer structures in the area of current flow, they can strongly reduce the measured currents.

Despite the aforementioned complications, it was possible to measure currents on self-assembled monolayers of molecules and bare Au surfaces with the presented CAFM setup, described in Chap. 4.

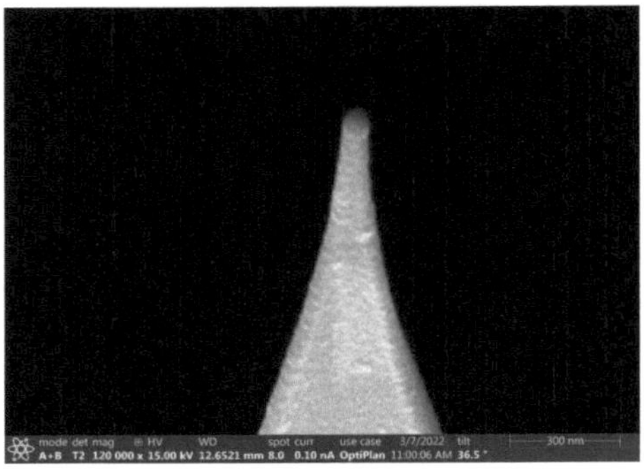

(a) Pristine.

(b) Damaged.

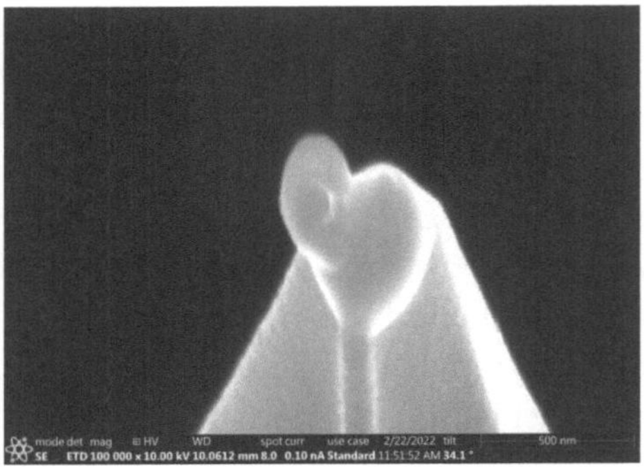

(c) Damaged.

Figure 3.6: SEM images of a pristine and damaged MESP-V2 probe apexes.

4 Electronic Properties of Molecular SAMs

The main aim of this work is to study a molecular motor, which is intended to be driven by the chiral-induced spin selectivity effect. To get a better understanding of the effect itself, studies on larger arrays of molecules were performed using the conductive atomic force microscopy (CAFM) setup presented in Chap. 3. This work was done in the group of Prof. Jan M. van Ruitenbeek at Leiden University in the Netherlands, in the framework of a Research Travel Grant provided by the Karlsruhe House of Young Scientists (KHYS).

The CAFM studies presented in this Chapter are mostly qualitative. To allow quantitative comparisons, significantly more experimental effort would be required. However, within series of measurements, also quantitative comparisons are appropriate. The conductance measurements were acquired with CoCr and solid Pt probes. As mentioned in Sec. 3.3, the force calibration for the solid Pt probes deviates significantly from the specified values. The CoCr probes suffer heavily reduced conductivity within less than $\sim 1\,\mathrm{week}$, also varying between probes, even from the same batch, rendering most quantitative results irreproducible with different probes and for different measurement series. This should be kept in mind when comparing forces and currents.

However, even without quantitative comparison, qualitative differences between substrates were observed in their conductive and structural properties. Moreover, statements about the measurability of currents through SAMs, especially for oligopeptides, can be made.

4.1 Motivation

The chiral-induced spin selectivity (CISS) effect, as already discussed in Subsec. 2.4.1, is a puzzling phenomenon to this day, where many studies have been performed and many theoretical descriptions have been put forward. Even though spin-dependent transport through molecules was observed in experiments, its origin still remains an open question and might even have multiple underlying effects, depending on details of the experiments.
To shed light on characteristics of CISS, in regard of the facilities available at Leiden University, the route of transport measurements is chosen, using CAFM. In this field, studies were already published, many from the group of Prof. Naaman at the Weizmann Institute of Science in Israel [58–61]. One particular study [41] is chosen as a starting point for the investigation of CISS in transport, in this case in self-assembled monolayers (SAMs) of oligopeptides on a Au substrate.
In this study, SAMs of helical oligopeptides, bound to a Au substrate *via* sulfur–gold bonds were studied using magnetic and conductive CoCr-coated AFM probes. More specifically, different currents were measured depending on the magnetization of the probe and the applied load force. They found that the currents through oligopeptide SAMs vary by a few pA depending on the probe magnetization and therefore spin-polarization of the injected current. Moreover, they found that the spin-polarization decreases with increasing load force.
The intention of the CAFM work in this thesis is to try to reproduce these results under improved conditions. The referenced publication [41] does not provide information on the type of Au substrate that was used or its roughness, neither is the inversion of the probe magnetization examined on a reference surface, which hampers the quantitative study of the dependence of the current on the spin polarization. Finally, the current measurements are averages over $50 - 70$ measurements at random tip positions. The rather small number of measurements does not allow for a discussion of parasitic effects other than the properties of the SAM and it is unclear whether $50 - 70$ measurements provide a sufficiently large statistical sample.
In the CAFM measurements presented here, reference measurements were made for the inversion of the probe magnetization, using the MFM signal of a strip of floppy disk tape (Sec. 4.2). Moreover, studies of the flatness of the used Au substrate and its conducting properties are presented in Sec. 4.3. As a further reference, the conductive properties of SAMs of achiral molecules, here dodecanethiols, were performed for Au/Mica and Au/Si substrates in Sec. 4.4, underlining the importance of the size of the substrate's terraces. Finally, Sec. 4.5 presents CAFM results on oligopeptide SAMs, where significant currents are only observed on small areas, even though the force seems to be high enough to remove the SAM.

Unfortunately, the given time frame of the Research Travel Grant did not allow for the reproduction of the exact results presented in [41]. However, qualitative statements about the importance of the substrate structure and the conductivity of SAMs of alcanethiols and oligopeptides are possible.

4.2 Inversion of Probe Magnetization

A first step to being able to measure the spin-dependent current through SAMs of helical molecules is to verify that the probe magnetization, and therefore the injected spin-polarization, can be inverted.
In [41], the orientation of probe magnetization and therefore spin-polarization of the current was achieved by placing the probe in a magnetic field of ± 1 T for 20 min. Here, to magnetize the probe in a certain direction, it was mounted as usual and was brought in contact with a strong permanent magnet by approaching until a certain, small force was applied. The probe was then retracted by ~ 3.5 µm to avoid wear of the probe apex and kept at that distance for 10 min. To invert the magnetization, the probe is approached onto the opposite side of the magnet.
To verify that this actually inverts the probe magnetization, a magnetic structure has to be imaged at both probe magnetizations, i.e. with MFM. If the contrast is inverted, also the probe magnetization is successfully inverted. For that purpose, a sample exhibiting magnetic contrast is needed, on which an exact location can be found after removing it from the sample stage, as the sample has to be replaced with the magnet for the inversion of the probe magnetization.
Also, the magnetic moment and coercivity of probe and sample have to be considered. If the magnetic moment of one is too large, and the other's coercivity is too small, the magnetic moments can align and no magnetic contrast is measured.
In this work, a lithographically gold patterned stripe of floppy disk tape serves as a platform to study the inversion of probe magnetization, as it meets the requirements stated above. It has a magnetic pattern inscribed, if it has data saved on it, which is detectable by MFM and the gold pattern allows to find the same part of the magnetic pattern that was imaged before probe magnetization inversion. Advantages of the floppy disk substrate are that it is readily and cheaply available and that its magnetic pattern is stable under the conditions of the gold pattern deposition.
The magnetic pattern of the floppy disk is arranged in stripes of ~ 200 µm width separated by stripes of no magnetic pattern. The magnetic pattern itself is a stripe pattern (perpendicular to stripes of magnetic/no magnetic pattern) of $\sim 5 \times 5$ µm^2 periodicity. A sketch is shown in Fig. 4.1a.
To ensure that the gold pattern and the magnetic pattern can be on one image, the gold pattern is chosen to be a 13×13 grid of 5 µm square markers on a total edge

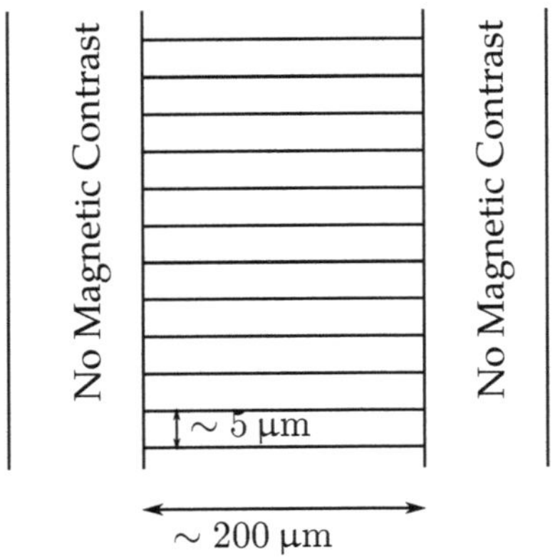

(**a**) Magnetic stripe pattern of the floppy disk substrate.

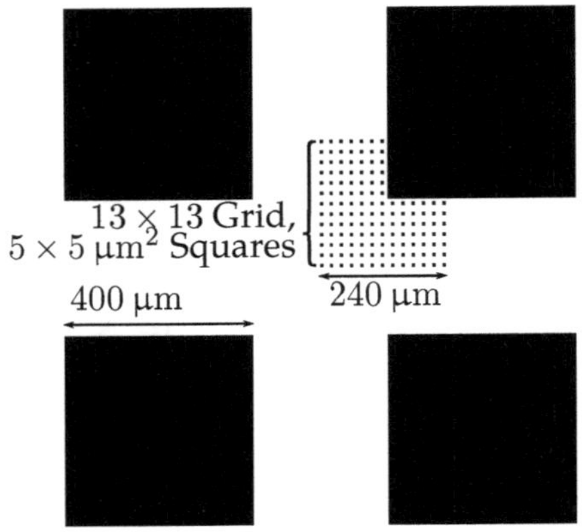

(**b**) Gold pattern deposited onto the floppy disk.

Figure 4.1: Magnetic pattern of the floppy disk and the gold pattern deposited onto it. The sizes are chosen, so that the 13 × 13 grid is on magnetic contrast, to find back the area measured after removing the sample to invert tip magnetization.

length of 240 µm, separating their centers by 20 µm. That way, the pattern fits the maximum range of the piezo scanner (30 µm), the single markers can be distinguished under the AFM microscope and they are on a large enough area so at least part of the gold pattern is deposited on an area with magnetic signal. For better visibility, the pattern is surrounded by four square markers of 400 µm edge length. The grid of markers is arranged asymmetrically in between the large markers to allow for easier distinguishability between the directions of the grid pattern. The complete pattern is sketched in Fig. 4.1b.
The patterning of the floppy disk was done completely at Leiden University facilities. It was spin-coated with polymethylmethacrylate (PMMA) and Electra 92[1], to dissipate charges due to the insulating sample, heating it up to 150 °C in the process. Using e-beam lithography, the pattern was written and subsequently developed. A ~ 30 nm thick gold layer was sputter-coated onto the surface with the PMMA layer still in place. In a lift-off process, the PMMA was dissolved in an acetone bath. The detached gold was removed from the surface by repeatedly sonicating the sample for short times, rinsing it with water, until only the written gold pattern remained. Indeed, imaging the patterned surface with AFM, the markers were visible and the sample stage of the AFM allows for a precise enough positioning to image the area around a chosen marker. An exemplary image of a marker on the floppy disk surface is shown in Fig. 4.2a. For some markers, very high structures are observed

[1] `https://www.allresist.com/portfolio-item/protective-coating-ar-pc-5090-02-electra-92/`

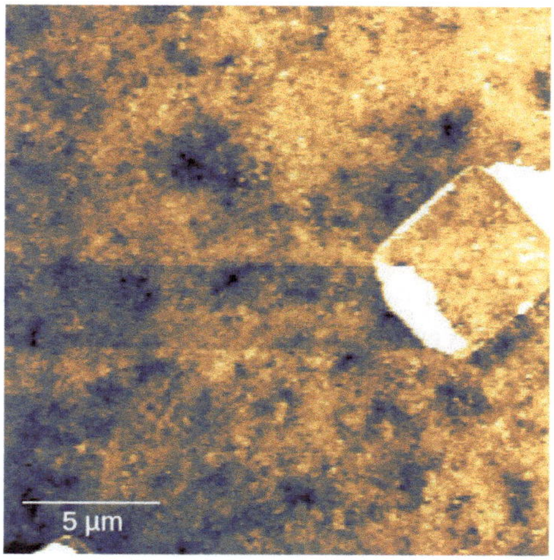

(a) Topographic image of the patterned floppy disk surface. The marker is clearly visible, both because of the height change and the overhanging gold layer.

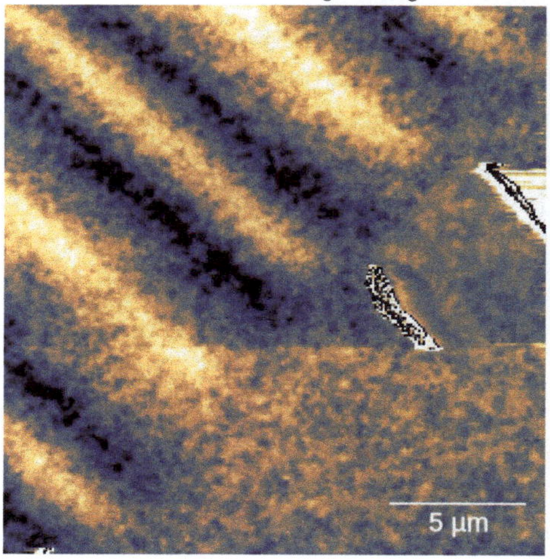

(b) MFM contrast at one probe magnetization.

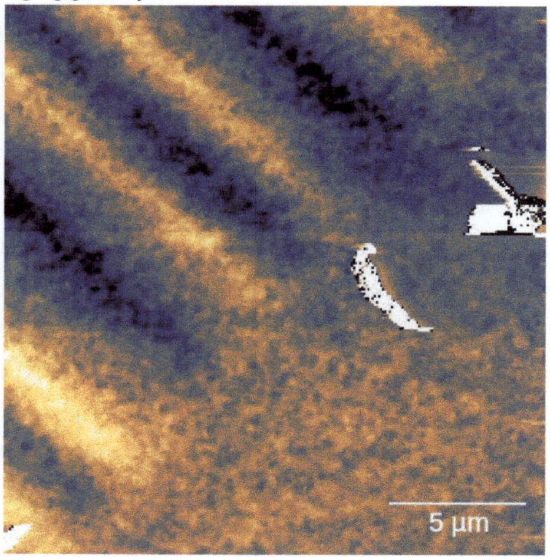

(c) MFM contrast for inverted probe magnetization.

Figure 4.2: Images of the patterned floppy disk. All images are taken on the same area. The MFM contrast is inverted for opposite probe magnetization, proving that the inversion procedure works (MESP-HM probe, AC mode, Hover at 100 nm distance for all images).

at the edges, which come from gold overhanging from the top of the marker due to rounded edges of the developed resist. These can serve as an indication of the marker in the MFM signal, as the probe is retracted much more than due to the topography of the sample if it gets into contact with the overhanging gold.
Two MFM images of the same marker for inverted tip magnetization, together with the magnetic contrast of the floppy disk are shown in Fig. 4.2b and 4.2c. Having the marker as a reference, the magnetic contrast can be compared between the two images, revealing that it is indeed inverted. It is the clearest for the slightly wider ares of magnetic contrast. Also, this process is repeatable and observed in three separate instances.
The inversion of magnetic contrast after the aforementioned procedure to invert the probe magnetization proves, that it works and can be used to inject currents of opposite spin orientation.

4.3 Au Surface Characterization

For the deposition of dodecanethiol and oligopeptide SAMs, a substrate has to be chosen. In the scope of this thesis, three different types of substrates were tested, whether they are suitable for CAFM studies on molecular SAMs. For that purpose, the surface structure and changes in conductivity were investigated. The tested substrates are template-stripped gold on mica, commercially available gold-coated silicon wafers, and gold on mica (Au/mica) substrates.
Template-stripping is a commonly used technique to obtain flat and clean gold surfaces [62, 63], with samples being commercially available[2]. The template-stripped gold used here was prepared at Leiden University by Jędrzej Tepper. The principle is to deposit Au on a freshly cleaved mica wafer in vacuum conditions, producing a clean interface, and gluing a small piece of mica onto the Au layer. If a fresh sample is needed, the glued mica piece is stripped from the mica wafer, revealing a clean Au surface. The flatness of the Au surface stems from the interface with the flat mica wafer.
An exemplary image of such a surface is shown in Fig. 4.3 (square fit to every line subtracted as background). On the complete 1 µm area, no deep indentations or very rough patches are observable. Some corrugation is visible ($\sim 2\,\mathrm{nm}$), which could represent impurities on the surface, defects within the Au, but close to the surface or terrace edges of the Au surface. Such a substrate could technically serve as an appropriate substrate for SAM formation. However, in previous experiments [64], it

[2] `https://www.platypustech.com/gold-thin-films/ultra-flat-gold-films`

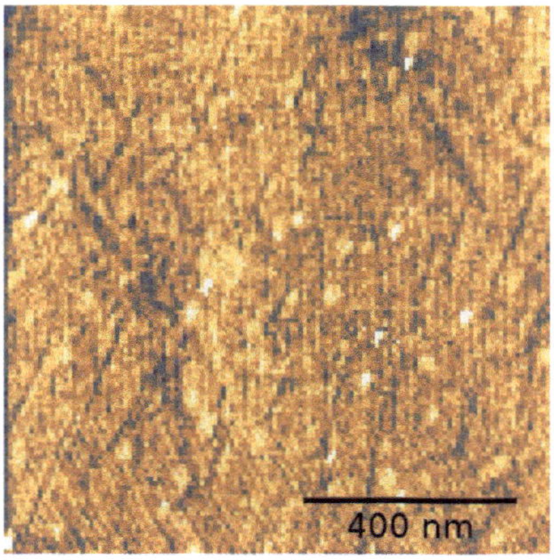

Figure 4.3: Image of template-stripped Au. On the complete 1 µm range, no deep indentations are observable. However, as SAM formation was not successful in previous studies, this substrate is not considered further. The corrugation is $\sim 2\,\text{nm}$ (MESP-HM probe, $F_{\text{set}} = 8\,\text{nN}$, $U = 0\,\text{V}$).

was not possible to achieve proper parallel alignment of dodecanethiol molecules in SAMs on such surfaces, possibly because the crystal structure on the gold surface is not well-defined. This type of substrate is therefore not considered for further investigations.
Another frequently used type of substrate for SAM deposition is a gold-coated Si wafer (Au/Si), also being commercially available[3]. The substrate is cleaned by boiling it in acetone and ethanol for 20 min under a fume hood, dried in a glovebox in N_2 atmosphere, exposed to ozone to remove organic contaminants and finally rinsed with warm ethanol. Figure 4.4a shows a topographic image of such a surface on the same scale as for the template-stripped Au in Fig. 4.3. The image shows a qualitatively grainier surface with only few areas where the Au surface is flat. Even though the corrugation is not too large ($\sim 7\,\text{nm}$), the grainy structure of the surface causes considerable height variation all across the sample.
As the images were measured in quantitative imaging (QI) mode, force-distance curves were measured at each pixel, also acquiring current-distance curves. Examples of such measurements are shown in Fig. 4.5. The force-distance curve is very similar to the expected curve (compare Subsec. 2.2.2). In this particular case, the result of too fast acquisition of the force- and current-distance curves at each pixel becomes evident. There is an overshoot in force in the retracting curve (light blue, overshoot visible from the force higher than the setpoint at the start of the retract curve), as the piezo is still expanding from the approach curve (dark blue), even though the retract curve is already being acquired. This also leads to a larger

[3]`https://www.sigmaaldrich.com/NL/en/product/aldrich/643262`

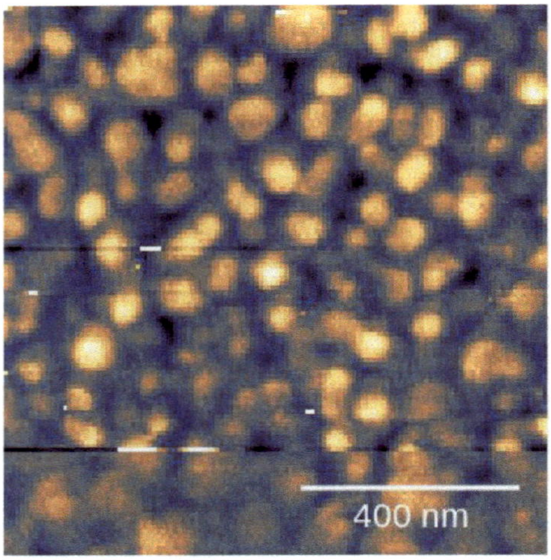

(a) Topographic image. It is comparably grainy and only has small areas which are flat. The corrugation is ~ 7 nm.

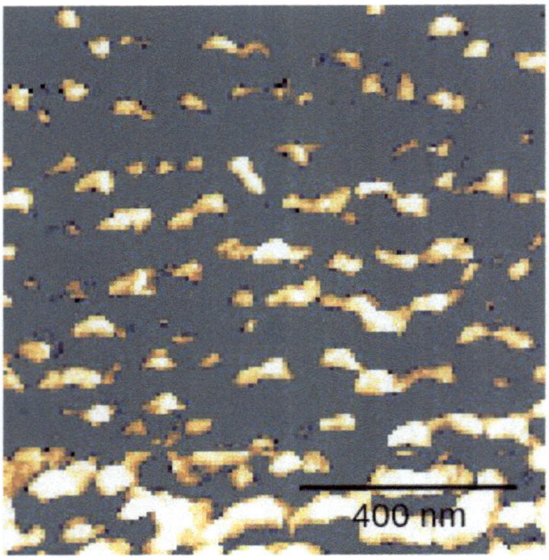

(b) Current map (log. scale). The extremal current varies strongly, being maximal in the areas between the flat areas (I_{min}=-7 pA, I_{max}=11.909 nA).

Figure 4.4: Topography and current map of an Au/Si surface. The image is the same size as Fig. 4.3 (RMN Pt probe, $F_{\text{set}} = 20$ nN, $U = 0.1$ V).

extremal current in the the current-distance curve in the retract curve (orange). To circumvent this problem, the extremal current is extracted from the retract curve for current maps.

A current map of the Au/Si surface is shown in Fig. 4.4b, in which the extremal current is plotted on a logarithmic scale for better visibility, which is done throughout the complete chapter. In these maps, blue and yellow areas represent low and high currents, respectively. The minimal measurable current is -7 pA, the current offset of the CAFM module (with added noise) and the maximum measurable current is the saturation current of 11.91 nA. This range is completely covered (with a few pA deviation) by all maps presented in this chapter.

The current map shows low currents for flat parts of the topography, both at high and low z values, and high currents only in areas where the tip is in between high, flat areas. This fact could be explained by different forces due to changed contact area at the edges or by varying conductivity at the probe apex and sides. For Au/Si, the contrast in the current map could be explained by the probe apex having a higher conductivity on its side, leading to a higher current in areas, where also the side of the probe contributes to the current. However, this is incompatible with data on the Au/mica substrate. Therefore, both mentioned explanations cannot fully explain the data. Although it remains unclear, why the high currents appear in areas with high slopes in z, the current varies from not measurable to saturation on small length scales.

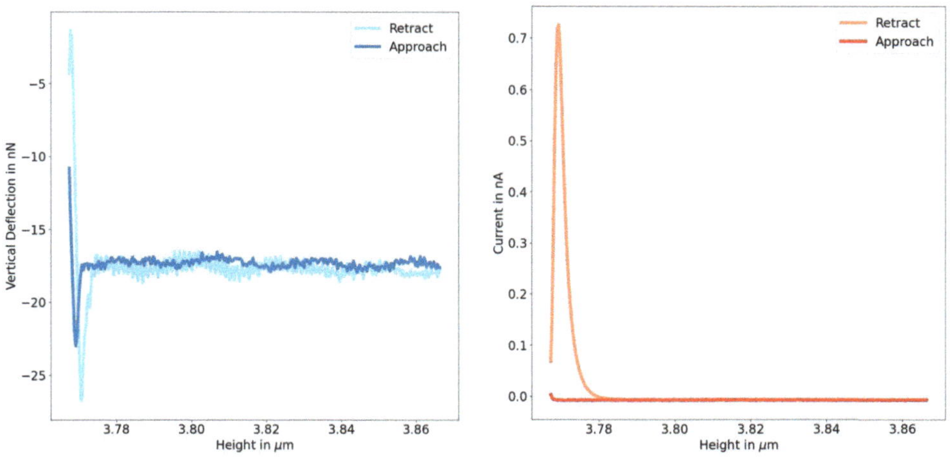

(**a**) Force-distance curve. There is an overshoot in force, as the retract curve starts at higher forces than the setpoint, due to fast scanning.

(**b**) Current-distance curve. The overshoot in force is also visible in the current, so the extremal currents are extracted from the retract curve.

Figure 4.5: Exemplary force-distance and current-distance curves at one pixel of a QI image. The dark colored curves are measured during approaching and the light curves during retracting of the probe (RMN Pt probe, $F_{\text{set}} = 6\,\text{nN}$, $U = 0.1\,\text{V}$).

A normalized histogram of the logarithmically plotted current is shown in Fig. 4.6a. This type of histogram plots the amount of pixels per unit current as a function of the current. The complete histogram is normalized to have an area of 1. In this case, the distribution only has significant values at very low currents (less than few $10\,\text{pA}$) and at saturation current ($\sim 12\,\text{nA}$). The fact that both these contributions do not yield meaningful data for the evaluation of the sample conductivity, in combination with the short length scales on which the current varies between the two extremes, shows that the sample does not have a suitable structure to meaningfully compare currents with spatial resolution.
Another possibility is that the conductivity of the probe plays a major role and influences the measured currents in such a strong way. This cannot be excluded, as the RMN probe sometimes lost its conductivity, showing that also significant changes can occur. However, as visible from Fig. 4.7b and as observed in similar experiments with the RMN probe, homogeneous conductivity can be achieved on Au/mica, but is not observed for Au/Si.
In the case of commercially available, epitaxially grown gold on mica substrates[4] (Au/mica), the situation improves significantly. A topographic image and current map on the same scale as for template-stripped gold and Au/Si is shown in Fig. 4.7.

[4] `https://www.phasis.ch/products/epitaxial-gold-au-111`

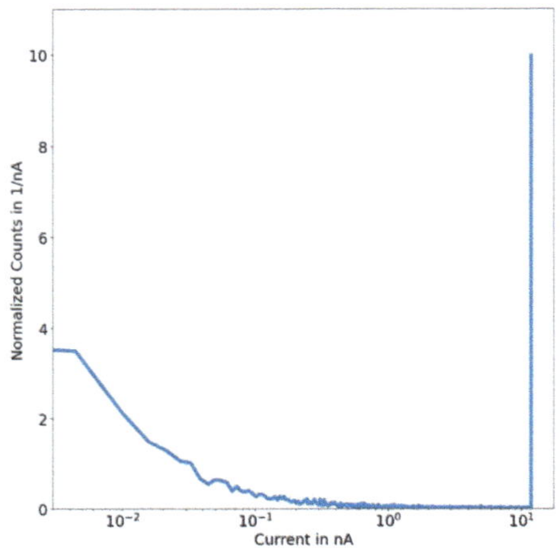

(a) Current distribution of the current map in Fig. 4.4b. The distribution only has a significant contribution at extremely low currents and saturation of the preamplifier.

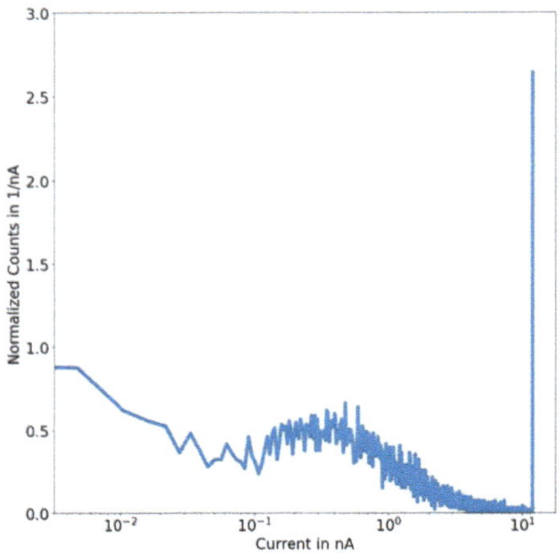

(b) Current distribution of Fig. 4.7b. Currents are distributed at the extreme ends of the distribution, but also at reasonable currents of $\sim 400\,\mathrm{pA}$.

Figure 4.6: Current distributions for Au/Si and Au/mica.

Even though the topography exhibits deep trenches, it also features large terraces. The corrugation is not smaller for the image ($\sim 9\,\mathrm{nm}$), but the large terraces provide large areas of well-defined Au surface, specifically with respect to its conductive properties.

The corresponding current map is also much clearer, exhibiting mostly homogeneous current on the terraces. For the trenches, the edges of terraces and the impurities, the current is not measurable, which is intuitive for the impurities, as they are usually organic contaminants that are insulating. For the terrace edges (and the trench), it could be explained by the contact area of the probe with the surface.

As the AFM only measures the deflection of the cantilever, it measures the force experienced by the complete probe. If the probe approaches an edge, it might get into contact with the higher lying terrace at the side of the probe, increasing the contact area. Therefore, the force that causes the deflection is effectively distributed over a larger area, reducing the force at each point. If the smaller force has a larger impact on the current, than the increased contact area, this could reduce the current. It should be noted that this argument is incompatible with the observations on Au/Si, where the current increases at the edges, assuming the probe shape and conductivity does not change between measurements.

The normalized histogram for the Au/mica substrate current map is shown in Fig. 4.6b. It shows a clear contribution of currents distributed around $\sim 400\,\mathrm{pA}$, which is in a measurable range for CAFM experiments, underlining the homogeneity of the currents on the flat Au terraces.

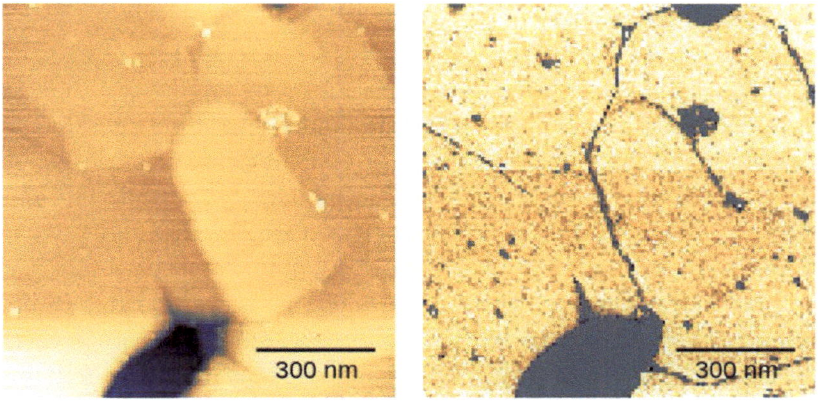

(a) Topography. Even though trenches are present, the flat areas are much larger than for Au/Si. The corrugation is $\sim$ 21 nm with the trench and $\sim$ 9 nm without.

(b) Current map. The current on the terraces is relatively homogeneous. Impurities on the surface and the trench exhibit low currents (I_{min}=−7 pA, I_{max}=11.909 nA).

Figure 4.7: Topography and current map of commercial Au/mica (RMN Pt probe, F_{set} = 6 nN, U = 0.1 V).

This indicates that the Au/mica surface yields qualitatively more homogeneous currents in a sensible range over large areas of the surface. This is beneficial in ambient CAFM studies of SAMs on gold, as large areas can be used to average over currents, necessitated by generally present, large fluctuations between measurements.

4.4 Dodecanethiol SAMs on Au

Using the commercial Au/mica substrate and the Au/Si substrate for comparison, self-assembled monolayers of achiral dodecanethiol molecules as reference were studied. Dodecanethiol (DDT) is an alkyl thiol, comprised of dodecane, which is terminated with a SH group ($H_3C–(CH_2)_{10}–CH_2SH$) and is a widely used model molecule for SAM formation [65–67]. As there is already experience in the group of Prof. van Ruitenbeek on how to achieve properly ordered SAMs of DDT on Au/mica, this molecule was chosen as a reference for the later investigation of oligopeptide SAMs. Here, it is also used to compare SAMs on Au/mica and Au/Si surfaces.
For the deposition, Au/mica substrate pieces from a freshly opened batch were immersed in a 10 mM solution of DDT in ethanol and incubated for 24 h. They were transferred into a glovebox, rinsed with ethanol and dried. For improved SAM formation, they were again immersed in a 10 mM DDT/ethanol solution, heated to $\sim$ 80 °C for 1 h, gradually cooled down and again dried in the glovebox. The deposition procedure is the same for Au/Si substrates, with the additional cleaning steps of the Au/Si substrates (described in Sec. 4.3).

4.4.1 Au/Mica Substrate

As presented in Sec. 4.3, commercial Au/mica substrates are flat and show homogeneous current distributions on terraces and are therefore suitable for SAM deposition and CAFM studies. A $300 \times 300\,\mathrm{nm}^2$ image of a flat area ($\sim 4\,\mathrm{nm}$ corrugation), with topography and simultaneously acquired current map, is shown in Fig. 4.8.
The topography shows smooth transitions between areas higher and lower in z. The features are visible, but rather blurry. The current map, however, shows features of the surface more clearly and reveals features, which are not visible in the topography. Firstly, the edges of terraces are much clearer and some edges are visible, which cannot be seen in the topography. Secondly, small areas ($\sim 5 - 9\,\mathrm{nm}$ diameter) of lower current, with no apparent periodicity or order, can be observed.
The corresponding current distribution is shown in Fig. 4.9 and is similar to the one observed for the Au/mica substrate, exhibiting a distribution of currents around $\sim 700\,\mathrm{pA}$, very well suitable to observe spatial variations of the conductivity in current maps. While there is a non-negligible contribution at saturation current, very low currents are rather rare.
Figure 4.10a shows a smaller-scale current map of the terrace at the left of Fig. 4.8b. The corresponding topography is basically featureless, but the small areas of reduced current are clearly visible in the current map.

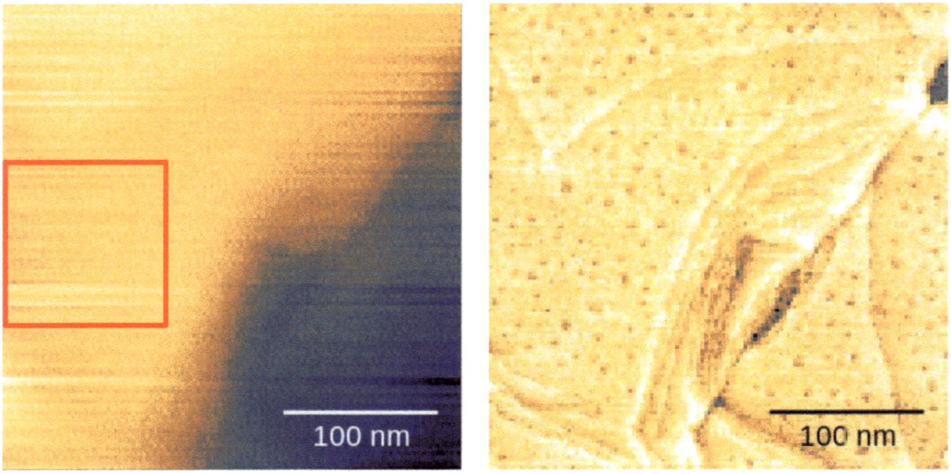

(a) Topography.　　(b) Current map (I_{min}=-7 pA, I_{max}=11.909 nA).

Figure 4.8: Topography and current map of a flat area of DDT/Au/mica. The topography shows smooth transitions, while the current map has sharp edges and shows additional features. The area of Fig. 4.10 is marked red. (MESP probe, $F_{\mathrm{set}} = 80\,\mathrm{nN}$, $U = 1\,\mathrm{V}$).

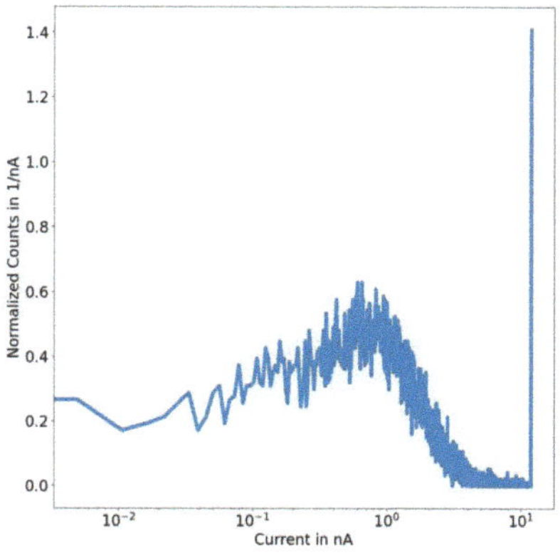

Figure 4.9: Current distribution of Fig. 4.8b. The currents are distributed around $\sim$ 700 pA allowing the observation of spatial variations in conductivity.

These areas are most likely so-called etch-pits due to SAM formation. As the DDT molecules bind to the surface *via* sulfur-gold bonds in large numbers, they lift the surface reconstruction (see Subsec. 5.4.1), which normally leads to a higher density of atoms on the surface than in the bulk. As the Au surface and the solution are in constant exchange, molecules bound to gold atoms can also be removed from the surface, in turn removing Au atoms from the surface. At first, this removes the excess Au atoms due to the lifting of the reconstruction. Then, it causes monoatomic depressions in the surface that nucleate and form vacancy islands [68], which can also be seen in STM [69, 70]. In the specific case presented here, these etch-pits could be observed in ambient STM on a sample prepared in the same way, also revealing that the molecules formed a well-ordered, standing phase [64].

Three subsequent current maps were acquired on the same area, shown in Fig. 4.10, demonstrating an evolution with every map. The current in the center increases, whereas it remains the same on the edges of the maps. However, the etch-pits are still present after the last map and seem unchanged. Note the visibility of the effect of drift in this series, being small enough to allow for comparison between the maps. The increase of current indicates improved conductivity of the surface. As DDT molecules are long, organic molecules they lead to reduced conductivity, compared to the bare Au surface. The increase in current is therefore indicative of a (at least partial) removal of molecules.

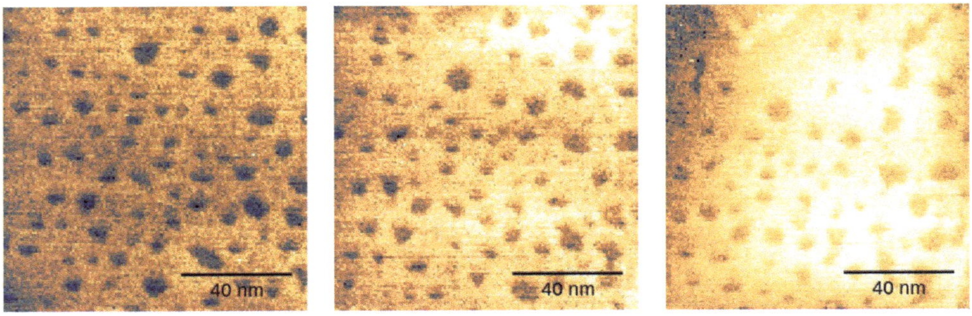

(a) Small-scale current map of terrace ($I_{\min}$=3 pA, $I_{\max}$=11.906 nA).

(b) Current map subsequent to Fig. 4.10a ($I_{\min}$=1 pA, $I_{\max}$=11.909 nA).

(c) Current map subsequent to Fig. 4.10b ($I_{\min}$=−6 pA, $I_{\max}$=11.909 nA).

Figure 4.10: Three subsequent current maps of the same area at the left hand side of Fig. 4.8b. Even though drift of the sample is visible, the maps are comparable. The current difference between the edges of the maps and the center gradually increases from Fig. 4.10a to 4.10c. The areas with lower current remain throughout the series (MESP probe, $F_{\text{set}} = 80\,\text{nN}$, $U = 1\,\text{V}$).

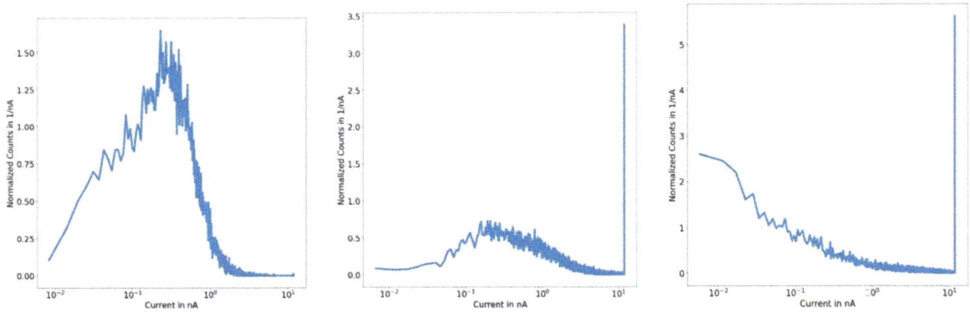

(a) For Fig. 4.10a.

(b) For Fig. 4.10b.

(c) For Fig. 4.10c.

Figure 4.11: Current distributions for the current maps shown in Fig. 4.10. The initial distribution has currents distributed at current values of $\sim$ 200 pA. During the removal, the distribution is shifted more and more towards the extremes (low currents and saturation).

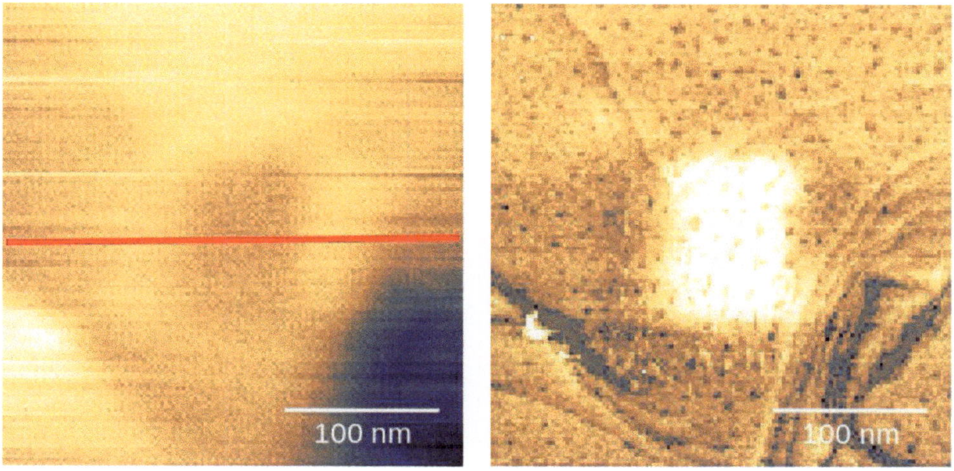

(**a**) Topography. (**b**) Current map ($I_{\min}$=−7 pA, $I_{\max}$=11.909 nA).

Figure 4.12: Topography and current map of the same area as in Fig. 4.8. The topography shows a clear indention on the area where the current maps in Fig. 4.10 are measured. The red line indicates the line along which the line profile in Fig. 4.13 is extracted. The corresponding current map shows increased current in the same area, indicating partial molecule removal (MESP probe, $F_{\text{set}} = 80\,\text{nN}$, $U = 1\,\text{V}$).

The current distributions for the current maps while partly removing the DDT SAM are shown in Fig. 4.11. While the currents in the first map are mainly distributed around $\sim 200\,\text{pA}$, the distribution shifts more and more to the extreme values (low current or saturation). The low currents are caused by the etch-pits and the high currents by the increased conductivity due to the removal of the SAM.
The suggested removal of molecules from the SAM is supported by the images shown in Fig. 4.12, presenting a larger-scale image of the same area, acquired after the three current maps in Fig. 4.10, where a change of the surface topography is clearly visible. It also shows that the current is increased on the area of the previous current maps, while retaining the etch-pits, which points towards a removal of the SAM.
However, measuring the profile of the indentation (see Fig. 4.13) in the topography yields a height difference of $\sim 450\,\text{pm}$, which is not large enough to be explained by a complete removal of the SAM, as the expected thickness of a standing phase DDT SAM on Au is $1.5\,\text{nm}$ [71]. This means that the DDT SAM is probably only partly removed and some molecules remain on the surface. Missing their neighbors to form a standing phase, the molecules arrange in a flatter manner on the surface. Still, it is also possible to measure current maps on this sample, without partly removing molecules from the imaged area, using lower forces.

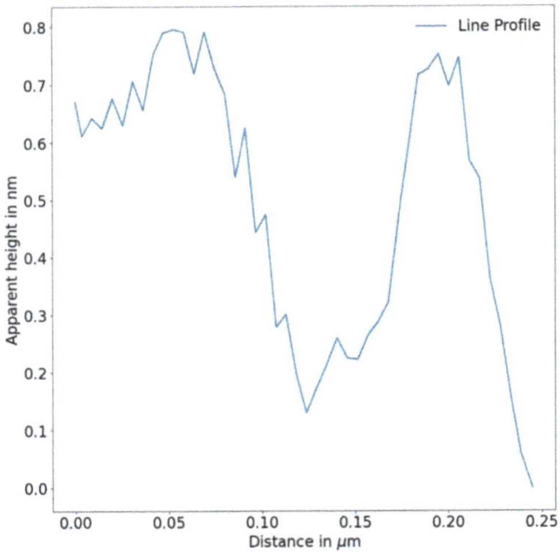

Figure 4.13: Height profile of the indented area seen in Fig. 4.12a. The profile is taken from a larger-scale image along the red line indicated in Fig. 4.12a at the same imaging conditions. The observed height difference is ~ 450 pm.

The reference measurements on DDT SAMs show, that it is important to choose a specific area, i.e. one terrace, to compare currents from CAFM measurements and averaging over measurements at e.g. a terrace and an edge might lead to large variations in current. Also, one needs to be aware of the forces that remove molecules of the SAM. If too high forces are applied, results might not be reproducible for subsequent current-distance or current-voltage spectroscopy measurements. Overall, substrates with large terraces are preferred for CAFM studies.

4.4.2 Au/Si Substrate

For comparison with DDT SAMs on Au/mica substrates, measurements on a commercial Au/Si sample were performed. Unfortunately, in the measurements on DDT/Au/Si, a prominent double tip feature (probe has two apexes that contribute in a similar fashion to force and current) was visible, which could not be removed. Despite this artifact, a qualitative comparison between the substrates is possible. For that purpose, topographic images and current maps of DDT/Au/Si are shown in Fig. 4.14. The images show topographies and current maps, the same size as previously shown for the Au/mica substrate (Figures 4.8a and 4.10a). It becomes apparent from the topographies and current maps, that the grainy surface of the Au/Si substrate (see Sec. 4.3) significantly reduces the areas of comparable currents. Only on the grains, comparing the measured currents is sensible.

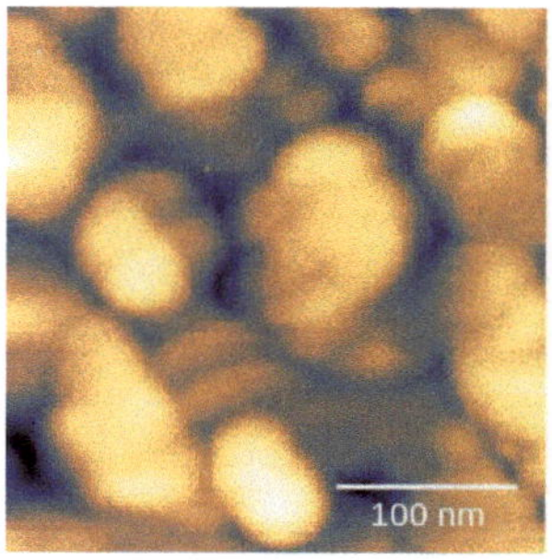

(a) 300 × 300 nm topography.

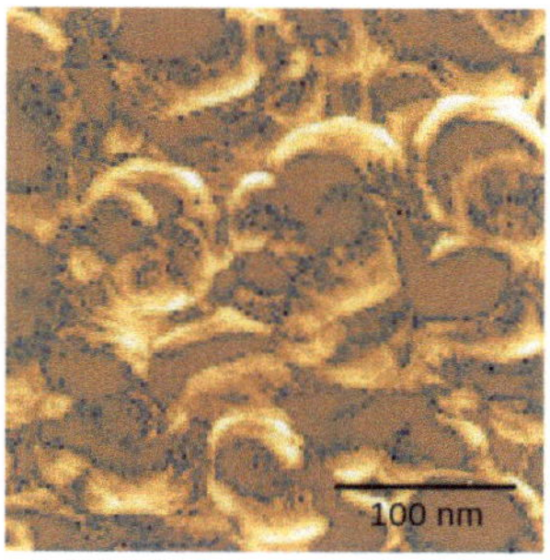

(b) 300 × 300 nm current map (I_{min}=−7 pA, I_{max}=11.909 nA).

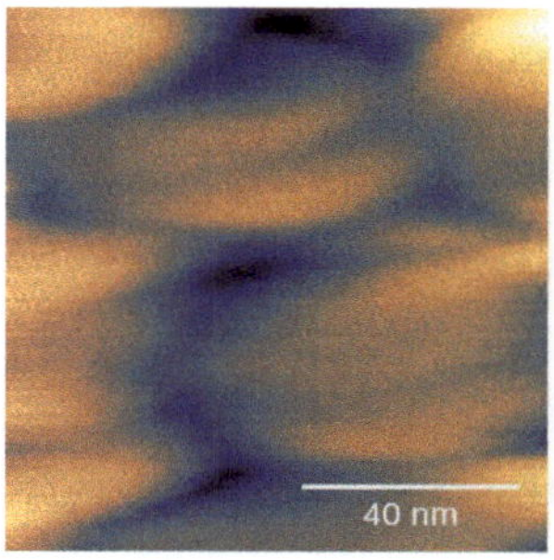

(c) 100 × 100 nm topography.

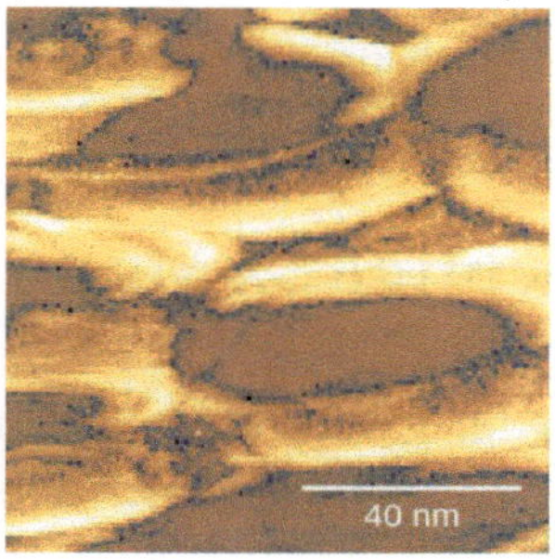

(d) 100 × 100 nm current map (I_{min}=−7 pA, I_{max}=11.909 nA).

Figure 4.14: Topographies and current maps of DDT/Au/Si at the same length scale as for DDT/Au/mica. Even considering the double tip feature, the grainy Au/Si surface significantly reduces the areas of comparable currents (RMN Pt probe, $F_{\mathrm{set}} = 20\,\mathrm{nN}$, $U = 1\,\mathrm{V}$).

When looking at the transitions between grains, the current becomes very large, which renders CAFM measurements challenging, as the high currents can lead to changes in the probe apex or coating, possibly melting the volatile probe apex (compare Sec. 3.3). In fact, in the measurements presented here, almost no current is measurable on the grains, compared to the areas in between. This behavior is most probably due to the dominating influence of the structure of the Au/Si substrate on the current, as discussed in Sec. 4.3.

The current distribution of the current map of DDT/Au/Si in Fig. 4.14d is shown in Fig. 4.15. It is similar to the one observed for the bare Au/Si surface. Almost all currents are below $\sim$ 100 pA. Above that, only saturation current is observed. Such conditions are not optimal for the study of spatial variations of conductivity.

The studies on DDT/Au/Si underline the importance of choosing a flat substrate with large terraces for CAFM studies on molecular SAMs, especially comparing the current distributions for the bare Au surfaces with the ones with a deposited SAM.

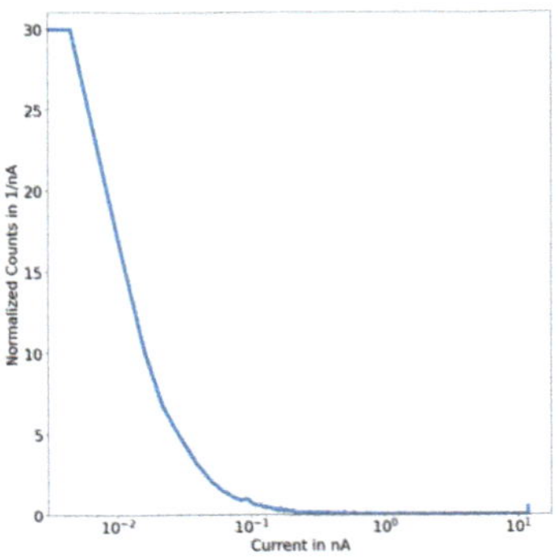

Figure 4.15: Current distribution for Fig. 4.14d. Almost only currents below ~ 100 pA or at saturation current are observed.

4.5 Oligopeptide SAMs on Au/Mica

With the knowledge gained from the studies presented on DDT SAMs, studies of oligopeptide SAMs can be performed. In this case, for comparability with [41], a commercially available[5] oligopeptide (SH–$(CH_2)_2$NH–(Ala–Aib)$_5$–COOH, called Pep5 in the following) was chosen, which forms a helical structure with a fixed handedness of the helix. As the commercially available Au/mica substrate appears to be the most suitable for CAFM studies, oligopeptide SAMs were deposited thereon. The SAM formation procedure was very similar to the formation of DDT SAMs. The Au/mica substrates from a freshly opened batch were immersed in a 0.1 mM solution of Pep5 in ethanol, incubated for 48 h and rinsed and dried in N_2 atmosphere.

A large-scale topography and corresponding current map are shown in Fig. 4.16. The topography shows a typical surface, as expected for the Au/mica substrate with some impurities on the surface and some displaced molecules (marked in the figure), although almost no current can be measured.

Again, smaller-scale images were acquired at intentionally high force and voltage to remove molecules from the surface, on the area, where molecules already seem to be displaced. The corresponding current maps are shown in Fig. 4.17. A similar behavior as for DDT SAMs is observed, where the current increases after each scan, again indicating the removal of the molecular SAM. Here, the area on which a current can be measured increases. This time, however, no etch-pits are observed and no current is measurable on parts of the current maps. This indicates that on the parts, where the molecules are not removed, the conductivity is reduced so strongly that no current could be measured, even at the large applied forces.

[5] `https://www.genemed.com/`

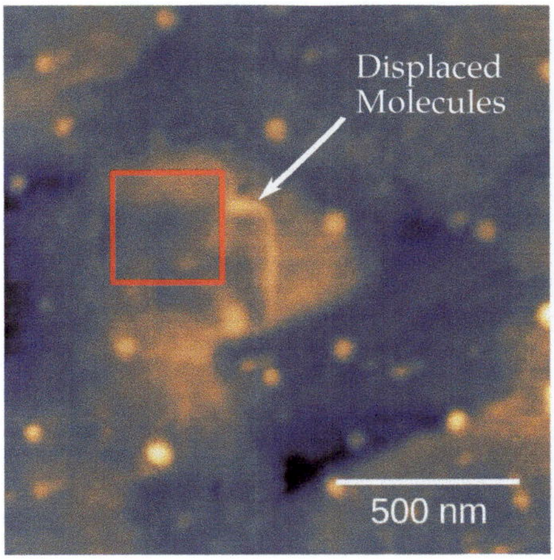

(a) Topography.

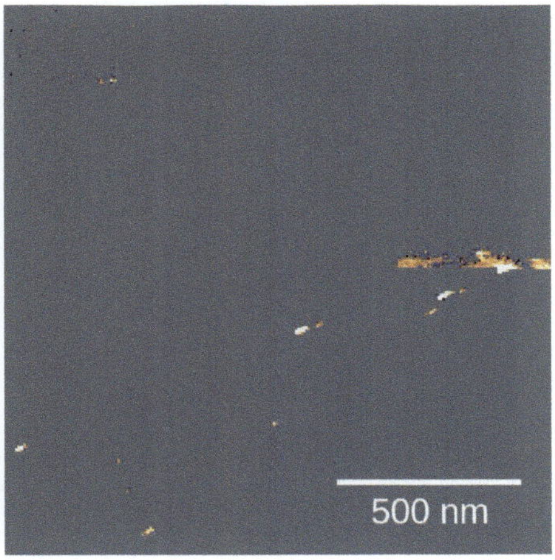

(b) Current map (I_{min}=−7 pA, I_{max}=11.908 nA).

Figure 4.16: Topography and current map of Pep5/Au/mica sample. The topography indicates that molecules are displaced, but no current is measured on the corresponding area. The area of Fig. 4.17a is marked red. Overall, no current can be measured on large parts of the current map (RMN Pt probe, F_{set} = 80 nN, U = 1.4 V).

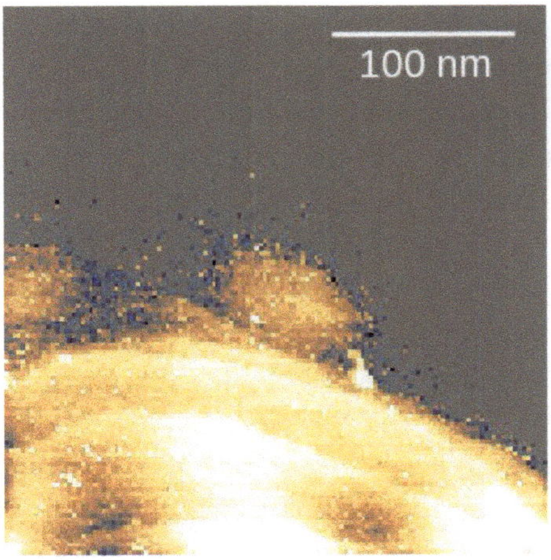

(a) Current map of the area marked red in Fig. 4.16a (I_{min} = −7 pA, I_{max}=11.909 nA).

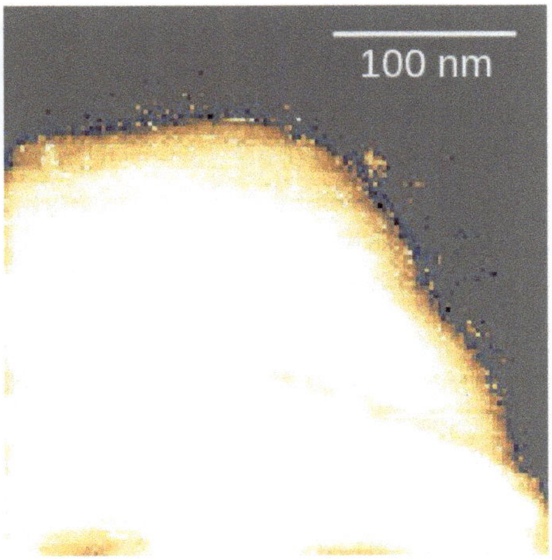

(b) Current map subsequent to Fig. 4.17a (I_{min}=−7 pA, I_{max}=11.91 nA).

Figure 4.17: Subsequent current maps. The images are intentionally acquired at large forces to remove the molecular layer. As expected, the areas with observable current and the current values increase. However, as long as the molecules are not removed, no current is measured (RMN Pt probe, F_{set} = 200 nN, U = 1.4 V).

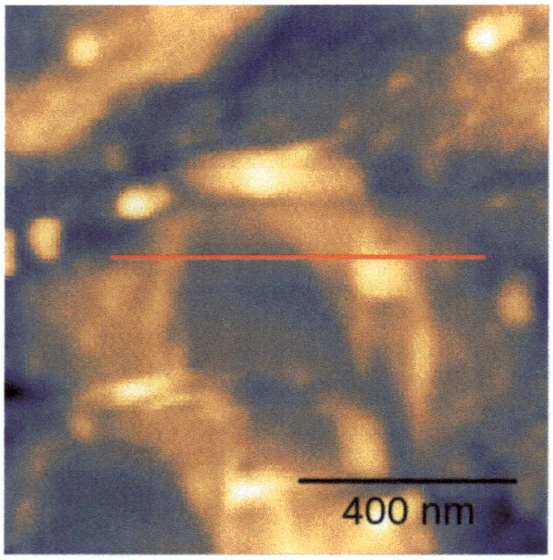

(a) Topography. The red line indicates, where the profile in Fig. 4.19 is taken.

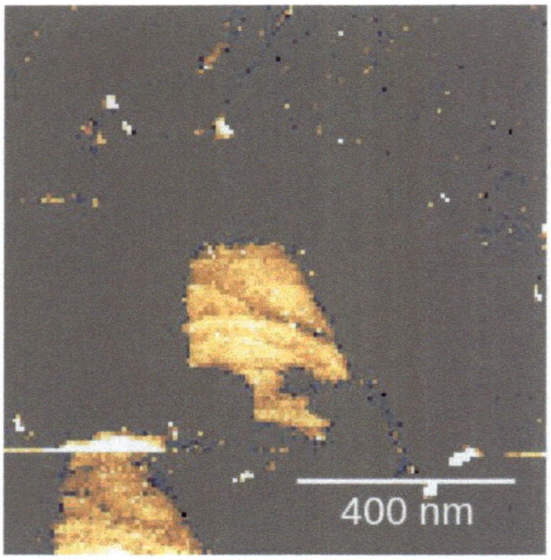

(b) Current map (I_{min}=−7 pA, I_{max}= 11.909 nA).

Figure 4.18: Topography and current map including the area of Fig. 4.16 after removing molecules. Despite the double tip feature, it is visible, that almost only the areas with removed molecules yield a measurable current (RMN Pt probe, $F_{set} = 60\,\text{nN}$, $U = 1.4\,\text{V}$).

Similar to the study of DDT SAMs, a larger-scale image, measured after molecule removal, is shown in Fig. 4.18. Unfortunately, a double tip feature was again present. Still, the topography reveals that the molecules are removed on the area of the previous images. Also, a current is only observed on these areas and on very small areas on the rest of the image, again pointing towards strongly reduced conductivity due to the Pep5 SAM. In fact, it seems to be reduced so strongly, that it might be challenging to measure a current without removing the molecular layer, as a current map could also not be acquired at an intermediate force of $F_{set} = 120\,\text{nN}$.

In analogy to the studies on DDT SAMs, a height profile is extracted from the topography and is shown in Fig. 4.19. Assuming that the flat parts next to the removed area represent the height of the surface with an intact SAM, the profile yields a height of the Pep5 SAM of $\sim 1.4\,\text{nm}$, which is in agreement with expectations [41]. This again indicates, that the complete molecular layer has to be removed to obtain measurable currents. Note, that the height measurement only holds, if the double tip feature does not influence the measured heights strongly.

In conclusion, the studies presented here do not allow for a detailed study of the chiral-induced spin selectivity (CISS) effect using a spin-polarized current from a magnetic probe in CAFM experiments. However, they emphasize the importance of choosing a substrate that provides large enough terraces of flat Au, to allow for measuring currents under comparable conditions for averaging. Moreover, the results on Pep5 SAMs underline how challenging the experimental observation of CISS using CAFM is.

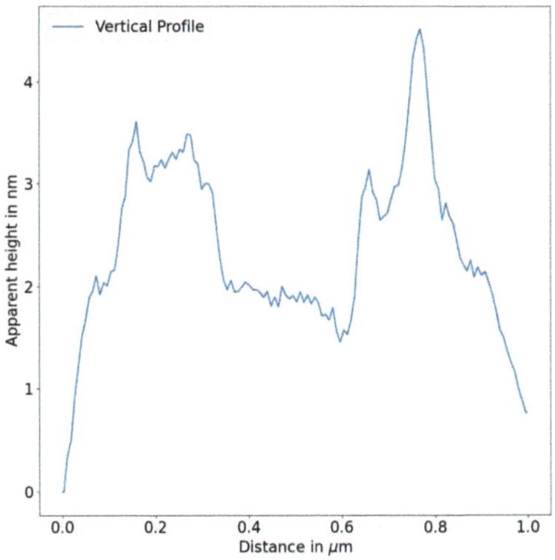

Figure 4.19: Height profile along the red line in Fig. 4.18a. A height difference of the surroundings and the area, where the SAM is removed of ~ 1.4 nm is visible, in agreement with the expected height of the Pep5 SAM.

5 Experimental STM Setup

This chapter is devoted to the experimental setup used to acquire the STM data examined and evaluated in the course of this thesis. First, an overview of the ultra high vacuum system is given. Next, the cryostat and the changes made to the thermal shielding of the STM and a description of the STM itself are presented. Finally, the properties of the used substrate, namely the (111)-surface of gold (Au(111)), together with the technique used to deposit molecules onto the surface are discussed.

5.1 Ultra High Vacuum System

Since highly clean samples and low temperatures are needed to perform the measurements, the STM has to be embedded into an ultra high vacuum (UHV) system. The low pressures prevent the samples from being polluted by adsorbates and from forming oxide layers on the surface. A sketch of the different chambers and pumps of the setup can be seen in Fig. 5.1. The rough vacuum is provided by a rotary pump. The chambers are furthermore pumped by turbomolecular pumps (TMPs) and ion getter pumps (preparation, analysis and STM chambers, not depicted in Fig. 5.1). This yields base pressures of the order of 10^{-9} mbar.
To be able to introduce new substrates and samples into the UHV system, it features the so-called loadlock. It has a rather small volume and a slot for standard *Omicron* sample holder plates. This way, it can be vented and pumped quickly, reducing time needed from the introduction of a sample into the UHV until first measurements. It is pumped by the TMP of the deposition chamber. The pressure in this chamber is not as low as in the others; common values are 10^{-6} mbar.

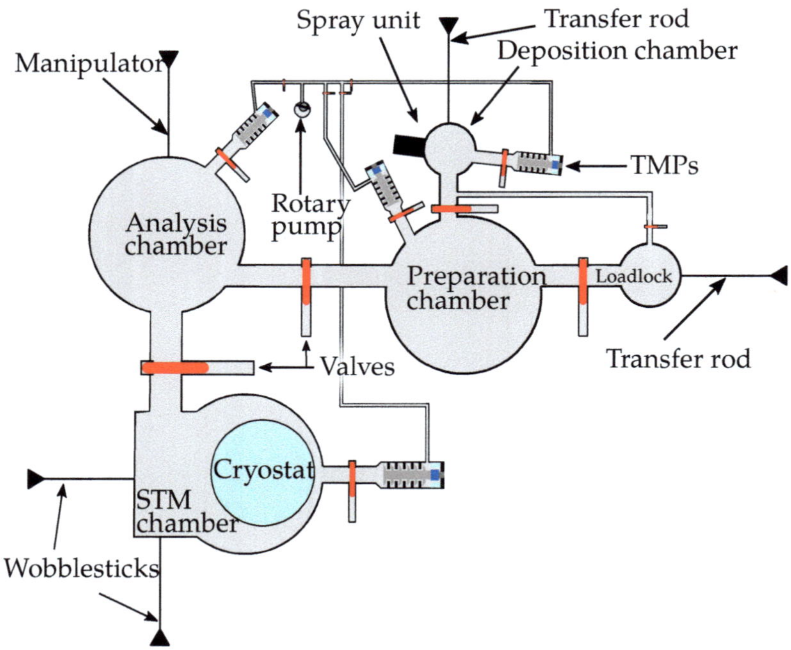

Figure 5.1: Sketch of the UHV setup with the STM chamber hosting the thermal radiation shields and the STM body, the analysis chamber for sputtering and annealing, the loadlock for transfer of samples in and out of the UHV and the deposition chamber and spray unit for molecule deposition. Not depicted are the ion getter pumps of the preparation, analysis and STM chambers.

From the loadlock, the sample is transferred into the preparation chamber, already at much lower pressures of 10^{-9} mbar. From there, it can either be transferred to the deposition chamber for spray deposition of molecules onto the surface, or to the analysis chamber. The spray deposition technique is described in Subsec. 5.4.2.
In the analysis chamber, the sample surface can be cleaned. Therefore, it provides a sputter gun with an attached argon supply and a manipulator for heating the sample. The cleaning process is described in more detail in Subsec. 5.4.1.
The clean sample is transferred to the STM chamber. It allows for sample storage in UHV for later use and holds the cryostat, the thermal radiation shields (Sec. 5.2) and the STM body (Sec. 5.3).
The complete UHV setup is located on four pneumatic vibration isolators from *Newport* [1] to reduce mechanical vibrations at low frequencies coupling in from the environment.

5.2 Cryostat and Thermal Shielding

All STM measurements were performed at low temperatures of $\sim 5\,\mathrm{K}$, thus freezing out degrees of freedom and improving energy resolution. These low temperatures are achieved by mounting the STM body on a liquid helium (LHe) bath cryostat at $\sim 4.2\,\mathrm{K}$.
The LHe cryostat is located inside a liquid nitrogen (LN_2) cryostat at $\sim 77\,\mathrm{K}$. This precooling step reduces the amount of LHe needed to keep the STM at $5\,\mathrm{K}$. It therefore leads to longer standing times without the need to refill LHe. This allows for longer and more stable measurements.
As the STM body is under UHV, the heat input through convection is rather small. The major heat input to the STM is via radiation. To minimize this heat input, the STM body is located within two thermal radiation shields, namely the LHe shield and the LN_2 shield. These are flanged and thermally coupled to the LHe and LN_2 baths, respectively.
In the course of this thesis, the thermal shielding has been changed, to allow for even longer standing times. While the previous thermal shielding was made of aluminum, the replacement was made from silver-plated copper. Moreover, it featured a mechanism for the STM to be positioned either pressed against a stamp protruding from the LHe cryostat, used e.g. to change samples, and a free-hanging position for improved vibration isolation and measurement, the so-called parking mechanism. However, this mechanism also formed a thermal bridge (although through weakly

[1] `https://www.newport.com/c/optical-table-supports`

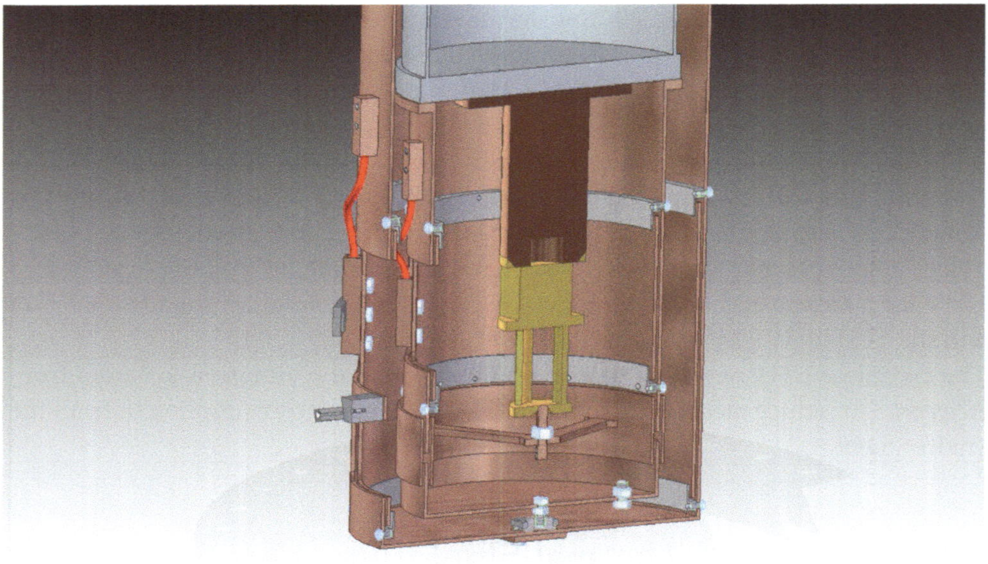

Figure 5.2: Cut through a technical CAD drawing of the new thermal shielding. The drawing includes the LHe cryostat (grey, top) and a mock-up of the STM body (yellow). The shields are made from copper and were silver-plated to reduce light absorption. The main difference is the parking mechanism (located below the STM body, pushing it against the stamp), which is now only attached to the LHe shield, reducing heat input to the STM, in turn increasing the standing time.

thermally conductive parts) to the outside of the LN_2 shield. This means that a part which is in contact with room temperature was also directly in contact with the STM body, causing an unwanted heat input.
The new shields are adapted from the design used in [72] and were mostly designed by K. Edelmann. However, in this case, no thermal shield for the Joule-Thomson stage is needed. Moreover, small adjustments are made to the parking mechanism. The parts were manufactured in the mechanical workshop of the Physikalisches Institut (PHI) at KIT. They are, just like in [72], made from silver-plated copper. The sanding, polishing, cleaning and silver-plating was done in-house. The main difference to the previous design is, that the parking mechanism is now solely attached to the LHe shield and is therefore also kept roughly at LHe temperature. This reduces the heat input to the STM significantly. The new thermal shielding (see Fig. 5.2) increased the standing time of the LHe cryostat from $\sim$ 5 days to more than 7 days.

5.3 STM Body

The STM body is the heart of the experimental setup. It features a holder for the sample and the STM tip. A false-colored CAD drawing is shown in Fig. 5.3.
As mentioned in Sec. 5.2, it can be positioned in the parked position, to enable sample transfer into the STM. In this position, it is pushed against a stamp, providing no vibration isolation but stability. In the unparked position, it is suspended from three copper-beryllium springs to reduce high-frequency vibrations being further reduced by two strong permanent magnets located at the sides of the STM body. These induce eddy currents in the STM body when it vibrates, causing a brake effect [73]. In this state measurements are performed.
For rough sample positioning, the sample slot is held within the body by stacks of piezo electric plates. These shear when a voltage is applied and with the slip-stick principle, the sample can be moved large distances compared to the single shear movement.
The tip is glued into a tip holder, which is fixed to a piezo tube opposite of the sample. The fixation of the tip holder was previously achieved by a spring on top of the piezo tube, pushing against the tip holder (as shown in Fig. 5.3). In the course of this thesis, the spring mechanism was replaced by a magnet fixed on top of the piezo tube, which allows for easier tip transfers and a more reliable fixation. As materials shrink to different extents upon lowering the temperature to 5 K, mechanical systems are rather unreliable and difficult to calibrate, whereas magnetic systems are comparably stable.
The complete magnetic system comprises sample plates with magnets to transfer the tip holder in (in-transfer magnet) and out (out-transfer magnet) of the STM and the magnet on the piezo tube. For the magnetic tip transfer to work, the strengths of the three magnets have to be tuned: That of the in-transfer magnet has to be the weakest, that of the piezo tube magnet has to be intermediate and that of the out-transfer magnet has to be the strongest. The tuning is achieved by heating the magnets above their Curie temperature for different amounts of time.
The piezo tube is used for the precise positioning of the tip. The piezo tube can extend and shrink such that it can move the tip in all spatial directions. Since deformations due to the piezoelectric effect are small, the positioning of the tip is very precise. The complete piezo tube is glued to the inside of a copper prism, which is again held inside the STM body by piezo stacks. This way, the tip can also be moved larger distances, again by slip-stick motion.

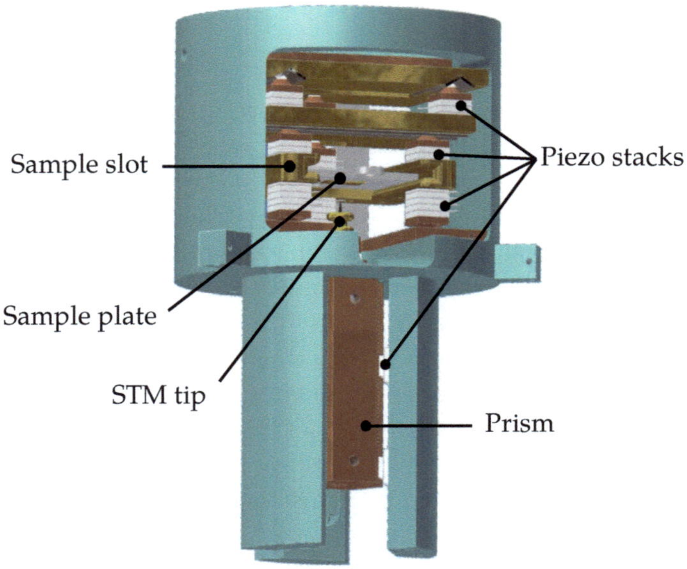

Figure 5.3: CAD drawing of the STM body. The sample fixed on a standard *Omicron* sample plate. Opposite of it, the tip is fixed by a spring to a piezo tube for precise tip positioning. The piezo tube is glued to the inside of a copper prism which can be moved by slip-stick motion of shear piezo stacks for longer-range movements along the z direction. The sample can be moved in both x and y directions laterally via slip-stick motion. Not shown are the springs and permanent magnets for vibration isolation and the thermometer.

All voltages needed for the movement of the piezo stacks and tube, as well as the bias voltage, are provided by a *Nanonis* system by *SPECS* [2]. The very small tunneling currents of a few pA are converted to measurable voltages through a commercial low-noise transimpedance amplifier from *Femto*[3]. A typical gain of such an amplifier is $10^9\,\frac{\mathrm{V}}{\mathrm{A}}$. This voltage is also measured by the *Nanonis* system. The commercially available software provided by *SPECS*, is used to set the necessary parameters for the measurements, such as the bias voltage between tip and sample, the feedback loop proportional gain and time constant, scan areas, bias spectroscopy settings, etc. and for all data acquisitions.

[2] `https://www.specs-group.com/nanonis/products/`

[3] `https://www.femto.de/en/products/overview.html`

5.4 Surface and Sample Preparation

As described in Sec. 5.1, the analysis chamber features a sputter gun and a filament for indirect heating of the sample. These are used to clean the surface of the gold single crystal which was used for all STM experiments. The gold crystal provides an extremely flat surface which is favorable due to the comparably small z range which can be covered by the piezo tube. The basic properties of the Au(111) surface, its cleaning process and the spray deposition technique for molecule deposition will be the topic of the following subsections.

5.4.1 Au(111) Surface

Gold is one of the chemically most inert noble metals [74]. Therefore, it does not oxidize and does not form strong bonds to many adsorbates. This makes it a very stable material, suitable for long, continuous measurements and depositions from solutions, which are not as clean as the sublimation of high-purity materials in UHV. Moreover, it is easily cleaned and handled, in order to provide a flat and clean substrate for molecule deposition. Another advantage is, that gold forms a relatively strong bond to sulfur, which can be used as anchoring of molecules to the surface [68, 75, 76].

Gold forms a face centered cubic lattice [77]. Therefore, the (111)-surface is a hexagonal lattice. However, the surface atoms form a $22 \times \sqrt{3}$ reconstruction [78]. Within the reconstruction, areas of hexagonally-close-packed (hcp) and areas of face centered cubic (fcc) arrangement are formed. This yields a higher local density of states at the transition areas between hcp and fcc stacking, which becomes visible in STM [79]. These areas are long, parallel stripes with edges, enclosing an angle of 120 °. An STM image is shown in Fig. 5.4. This characteristic structure is the so-called 'Herringbone' reconstruction.

In order to obtain such flat and clean surfaces, the Au(111) crystal was repeatedly sputtered and annealed. Sputtering is a bombardment with Ar ions, which removes atoms from the surface. This leaves behind a rough surface. To reduce the roughness of the surface, the sample was annealed. This means that it was indirectly heated to $\sim 500\,^{\circ}$C promoting diffusion and creating a flat surface. The molecules were then deposited onto the Au(111) crystal.

Figure 5.4: STM image of the Au(111) surface showing the Herringbone reconstruction and a step edge.

5.4.2 Spray Deposition

A clean and convenient way of depositing molecules on a surface is the evaporation of high-purity powder from e.g. a Knudsen cell from within the UHV system. However, the molecules studied in this thesis are not easily sublimed. If one were to evaporate them, a large fraction would disintegrate, leaving only fragments on the surface. Therefore, a spray deposition technique is used.
A sketch of the spray unit (compare Fig. 5.1) can be found in Fig. 5.5. The gold crystal is placed on a transfer arm in the center of the deposition chamber. The spray unit is a long pipe with a flange with a hole at the end. On the outside the chamber on that flange, a piece of magnetic tape is glued on top of the hole, having itself a hole at the same position. For the deposition, a piece of magnetic steel foil with a small pinched hole is placed on the magnetic tape. The spray unit can be evacuated with an oil-free rotary pump. A plate valve protects the deposition chamber from being vented through the spray unit, which is pumped by a TMP.
For deposition, the molecules were dissolved in dichloromethane (DCM), which had previously proven to be a suitable solvent [80]. A glass pipette was dipped into the solution. The capillary forces suck a small amount of solution into the pipette. It was then placed on the steel foil. The hole in the foil should have roughly the same diameter as the inner diameter of the pipette, to ensure good sealing of the connection without too much overlap with the steel foil.
Due to the pressure difference between atmospheric pressure behind the pipette and $\sim 10^{-3}$ mbar in the - now connected - deposition chamber and spray unit, the solution is sucked into the chamber. It rapidly expands and small droplets of solution land on the Au(111) surface.

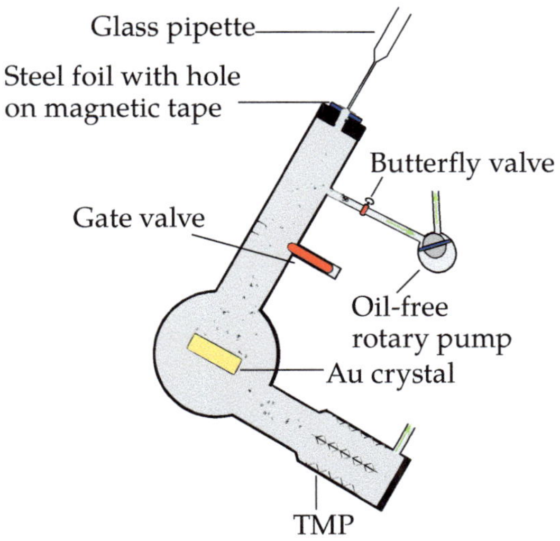

Figure 5.5: Side view of the spray unit attached to the deposition chamber. The gold crystal is positioned opposite of a piece of steel foil with small hole pinched into it. A glass pipette with the molecules in solution is placed on the hole. Through the pressure difference, the solution is sucked into the chamber and sprayed onto the gold surface.

After deposition, the crystal is post-annealed at $\sim 150\,°C$. This increases diffusion of the molecules and therefore the formation of ordered islands. Moreover, as it is also described in Subsec. 6.1.1, it promotes the cleaving of the acetyl groups in the thioacetates of the foot structures. This leaves the bare sulfur atom for the formation of sulfur-gold bonds to fix the molecules on the surface.

This concludes the description of the experimental setup. The following chapter will describe preliminary studies on molecules with two different, smaller tripodal platforms.

6 Adsorption and Self-Assembly of Tripodal Molecules

This thesis focuses mainly on single molecules consisting of a tripodal platform and a functional head group. This chapter presents different approaches to be able to address the functional group of such molecules with the STM tip, while it is still being free to rotate. This includes smaller platforms in Sec. 6.1 and an extended platform with a variety of head groups in Sec. 6.2. All molecules are specifically designed and synthesized for this project by Nico Balzer and Michal Valášek from the group of Prof. Marcel Mayor at the Institute of Nanotechnology (INT) at KIT. All studies of this chapter are aimed at finding a system which can support a chiral, rotatable head group to point away from the surface and be studied on a single molecule level. It is crucial to investigate a system which provides sufficiently reproducible conditions to study the rotational behavior of the chiral head group in detail. At the end of this chapter it will become evident, that a codeposition of two molecules provides suitable conditions to address the head group in STM experiments.

6.1 Tetraphenylmethane-based Platforms Tpd and sTpd

This section reports on results obtained on molecules with smaller tripodal platforms as carriers for functional groups. The first molecule presented is a tetraphenylmethane (TPM) derivative with a chiral head group, for which the head group is shown to adsorb on the surface, rendering it unsuitable for the intended study of the rotation of only its head group.
Subsection 6.1.2 presents a spirobifluorene scaffold, which is more rigid, as a potential platform to support a chiral head group. Results from ordered islands and single molecules show, that it can support a smaller head group when isolated, but will again adsorb with the head group on the surface, if arranged in an ordered island. The studies on these two molecules provide evidence for the necessity of using an extended platform to let the chiral functional group point away from the surface.

6.1.1 Tpd-G: Geländer Functional Group

The design of Tpd-G is based on molecules previously studied in this research group. The molecules are TPM derivatives with three thioacetate (SAc) groups attached to three of the benzene rings. The fourth benzene is replaced by a functional group. In the post-annealing process the acetate groups are cleaved off, deprotecting the sulfur atoms which can then bind to the gold surface. This immobilizes the molecules on the surface. The idea is that the three benzene rings, bound by sulfur-gold bonds, form a tripodal (Tpd) platform and force the functional group attached to the central carbon atom to point away from the surface. In previous studies done in this research group the functional group was chosen to be a benzene ring with a cyanine group [81] or elongated groups with a number of benzene rings and a cyanine termination [82].
As logic continuation, the same platform was chosen to mount the chiral functional group. In this case, it is a so-called 'Geländer' group (G), resulting in the chemical structure of Tpd-G shown in Fig. 6.1a. It is comprised of two benzene rings, which are connected via a link. This link consists of three carbon atoms with a double-bonded oxygen at the center. The linker between the benzene rings is too long to form a planar structure which leads to a tilt angle between the two benzene rings. This causes the chirality of the head group. In this case, the SAc groups were chemically deprotected, meaning that the Ac groups were replaced by hydrogen atoms, which can be cleaved off more easily.

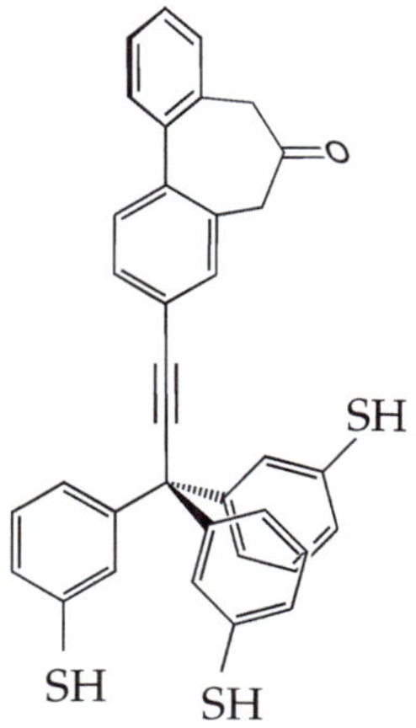

(a) Chemical structure of Tpd-G. The functional group is chiral due to the link between the two benzene rings being too long to form a planar structure, creating a tilt angle between them.

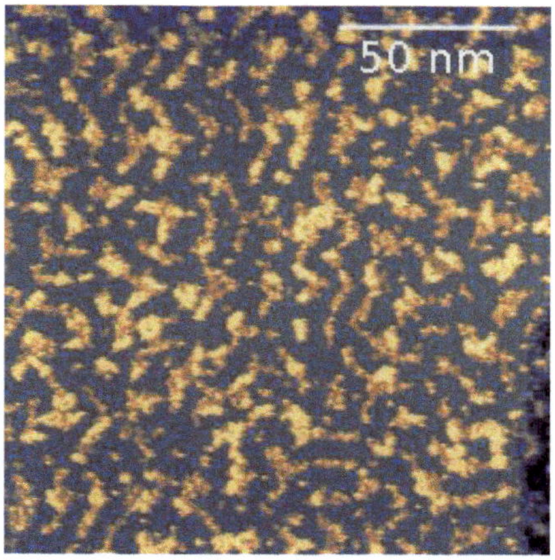

(b) Large-scale image showing that there is no formation of ordered structures, even though the coverage is high enough. The apparent height ($\sim$ 400 pm) is comparable with previous studies, evidencing that the molecules do not stand upright ($U = 0.4\,\mathrm{V}$, $I_{\mathrm{set}} = 9\,\mathrm{pA}$).

Figure 6.1: Chemical structure and deposition of Tpd-G on Au(111).

A larger-scale image of Tpd-G on Au(111) is shown in Fig. 6.1b. In the image, no ordered structures can be observed on the surface. Under similar preparation conditions, such ordered structures do form with similar molecules (see e.g. [82]). This points to the fact that the functional groups play a crucial role in preventing the formation of ordered islands. This might be due to the asymmetry of the structure or an unfavorable charge distribution within the molecule.
The apparent height of the molecules on the surface is $\sim 400\,\mathrm{pm}$. This is very similar to the apparent height observed for similar molecules, if one of the feet of the Tpd platform is pointing away from the surface [82]. This indicates that these molecules are not adsorbed with all three feet attached to the substrate. A more probable scenario is that the functional group is adsorbed on the surface. The objects lower in apparent height ($\sim 250\,\mathrm{pm}$) are most probably remnants of the solvent, impurities introduced by the deposition procedure and disintegrated molecules [83].
More interestingly, when looking at areas with lower coverage, circular structures could be observed. An exemplary image is shown in Fig. 6.2a. Their apparent height is similar to the unordered islands mentioned previously, again indicating that the molecules adsorb with the functional group on the surface.
The radii of the circles vary, but seem to be distributed around $\sim 2\,\mathrm{nm}$ (see Fig. 6.3). Even considering the broadening of sharp objects by the shape of the tip, these radii are much too large for the molecule to stand upright. This is another indication that the molecules are lying on the surface.

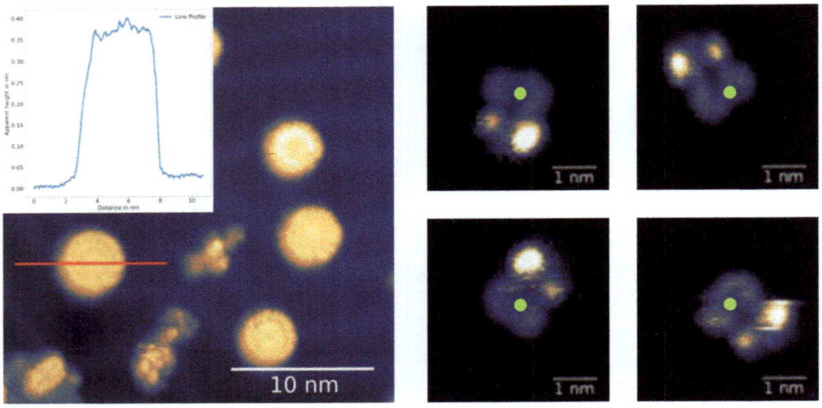

(**a**) Circular structures observed. They have a similar apparent height as the unordered structures in Fig. 6.1b, have sharp edges within and their diameters vary (see Fig. 6.3). The inset shows the profile along the red line ($U = 1.5\,\mathrm{V}$, $I_{\mathrm{set}} = 8\,\mathrm{pA}$).

(**b**) Four metastable states of Tpd-G. The images are taken subsequently on the same molecule in constant height mode. The structure stays the same but is rotated around a pivot point roughly in the center of the structure (green dot) ($U = 0.4\,\mathrm{V}$, $I_{\mathrm{max}} = 600\,\mathrm{pA}$).

Figure 6.2: Rotation of Tpd-G.

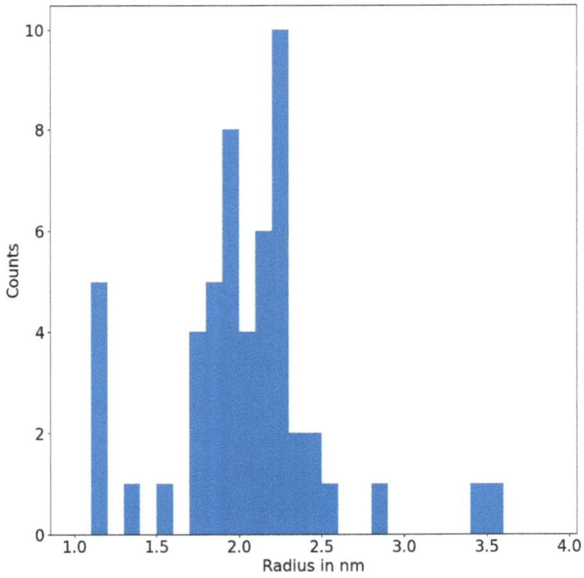

Figure 6.3: Histogram of observed radii of lying Tpd-G molecules. The distribution is centered around $\sim 2\,\mathrm{nm}$.

Upon closer inspection, the circular structures exhibit sharp edges, radially extending from the center to the rim of the circle. These are interpreted as different states during a rotation of the molecule around one fixed point. The molecule is only fixed to the surface at the center of the circular structure. During the scan, it is moved by the interaction with the tip, passing metastable adsorption states. The transition from one metastable state to the next is fast compared to the scan speed. It changes the apparent height instantaneously and therefore creates sharp edges in the circular structure.
This can also explain the different radii of the circles. If the point varies, at which the molecule is fixed, the maximum distance from the center changes, altering the radius of the circle.
Assuming the above, it could be possible to image the molecule in the different metastable states. Indeed, in one instance, a series of four constant height images could be taken, showing four of the metastable states, shown in Fig. 6.2b. Clearly, the visible shape stays the same, but is rotated around a fixed point (green dot in Fig. 6.2b). It was even possible to see a rotation mid scan, where the image showed an abrupt change from one line to the next. Combining the observations provides rather clear evidence for the interpretation that the molecules are attached to the surface by the functional group.
This means that the measurements on Tpd-G presented here were performed on lying molecules. In this configuration, it is highly unlikely that a rotation with a preferred direction can be observed, since there is no driving force to be expected. The Tpd platform does not seem suitable to carry the Geländer head group. Another platform has to be chosen for lifting the Geländer group from the surface, enabling the observation of a rotation with preferred direction. To solve this problem, a more rigid platform, e.g. a spirobifluorene platform, is one approach.

6.1.2 sTpd-L: Spirobifluorene Longneck

As concluded from the studies on Tpd-G, it is necessary to make adjustments to the tripodal platform, ascertaining a head group as large as the Geländer head group pointing away from the surface. A possible way to achieve this is to use a more rigid platform. The spirobifluorene tripodal (sTpd) platform was chosen, because this approach has proven successful [84].
The sTpd platform with a benzene ring attached via triple-bonded carbons and terminated by a cyanide group is shown in Fig. 6.4a. This structure allows the functional group to point perpendicular to the plane defined by the three SAc groups, giving it the name 'Upright Longneck' (L). What makes the sTpd more rigid are two interconnections (marked red in Fig. 6.4a). They reduce the number of degrees of freedom for each benzene ring. In turn, this reduces the number of possible

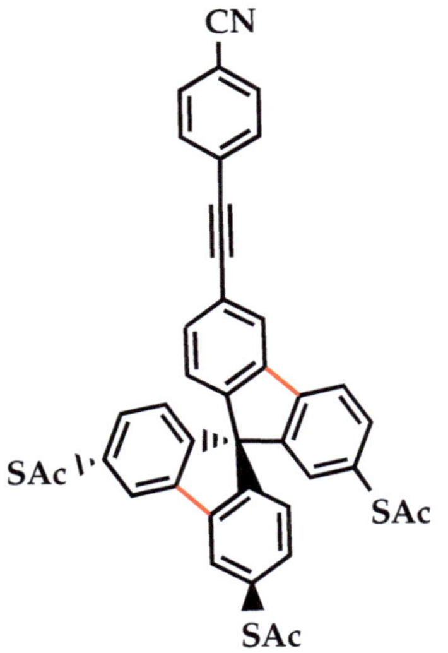

(**a**) sTpd-L chemical structure. The interconnections (red) make the sTpd more rigid, potentially allowing the functional group to point away from the surface.

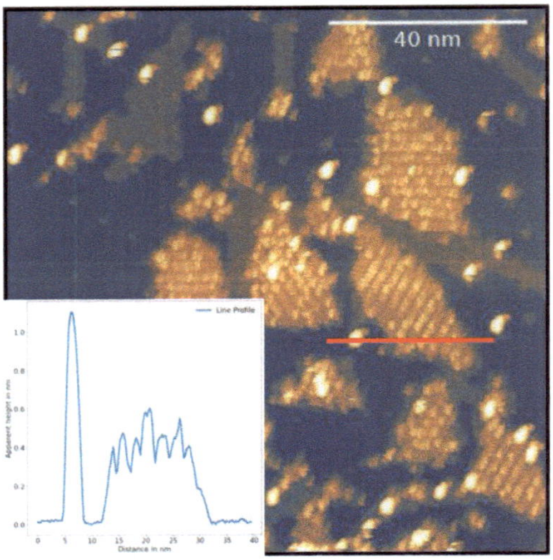

(**b**) Overview of structures on the surface. One can distinguish two features, the shallower ordered islands and the higher single molecules (bright dots). The inset shows the profile along the red line ($U = 2$ V, $I_{set} = 11$ pA).

Figure 6.4: sTpd-L chemical structure and features on the Au(111) surface.

configurations for adsorption on the surface, making it more likely that all three thiol-gold bonds form.

A typical scan of the surface is shown in Fig. 6.4b. There are two clearly distinguishable features. Firstly, there are ordered islands of ~ 400 pm apparent height. Secondly, there are comparably high structures of > 1 nm apparent height. They appear mostly separate or at the rim of clusters, rarely being incorporated in ordered islands.

The apparent height of the ordered islands suggests that the molecules forming them are adsorbed with the functional group lying on the surface. An image showing the periodicity of the ordered structure can be found in Fig. 6.5a. The unit cell has a size of 2.1 nm^2 and contains two molecules. Therefore, the footprint of one molecules is 1.05 nm^2. This is comparable to previous studies on molecules of similar size [82] which were also adsorbed with the functional group on the surface.

Figure 6.5b shows the proposed order of the molecules within the islands. The longer functional groups point in the direction of larger distance between the rows. In each row of the island, the partial charge in the molecule, caused by the cyanide group, is aligned anti-parallel. This possibly leads to an attractive electrostatic force, stabilizing the structure. The foot structure shows the same behavior as observed for the Tpd platform. The center molecules appear slightly higher and almost coalesce, while the molecules at the edge are flatter and appear round and separated. This can

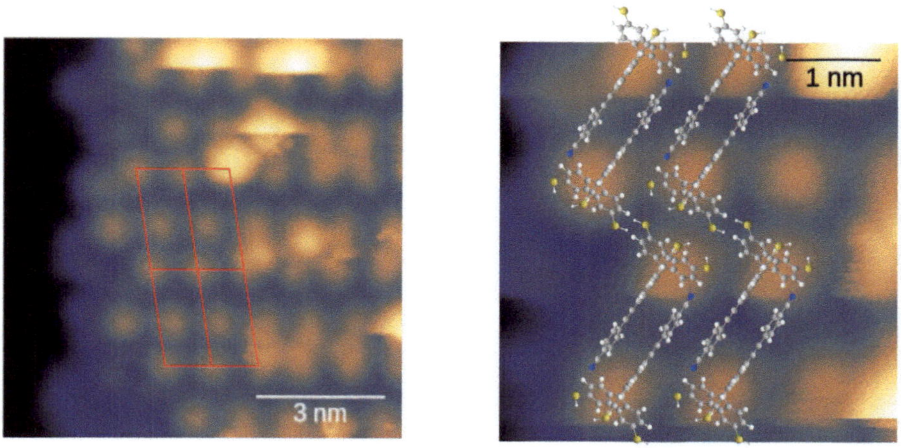

(a) Edge of an ordered island. The unit cell is marked with a red parallelogram. The molecular footprint matches previous studies.

(b) To-scale superposition of lying sTpd-L. The anti-parallel alignment of the cyanide groups supports the arrangement.

Figure 6.5: Ordered islands of sTpd-L (both at $U = 2\,\mathrm{V}$, $I_{\mathrm{set}} = 16\,\mathrm{pA}$).

be understood as the molecules within the island still carrying the acetyl protection group, while the molecules at the edges, where they can move more freely, lost all three protection groups [82].

In contrast to the ordered islands, single molecules of sTpd-L appear much higher, indicating that the head group is pointing away from the surface. Figures 6.6a,b show images of the same sTpd-L molecule imaged at bias voltages of $+2\,\mathrm{V}$ and $-2\,\mathrm{V}$, respectively. The blue and orange lines indicate the lines along which the cross-sections in Fig. 6.6c are taken.

As can be seen, the molecules appears wider when imaged at positive bias voltage. This can be explained by the electrostatic interaction between tip and the dipole of the cyanide group of sTpd-L. The bias is applied to the sample, meaning that at the tip is negatively charged for positive bias. This creates a field in the STM junction which is aligned anti-parallel to the dipole of the cyanide group, creating an attractive interaction and making the molecule appear wider. The inverse situation is the case for the opposite bias voltage.

The different size of the molecule could also be explained by tunneling into different orbitals. However, the sharp change form one line to the next at negative bias and the ripples in the scan at positive bias cannot. All these observations indicate that the functional group is indeed pointing towards the tip.

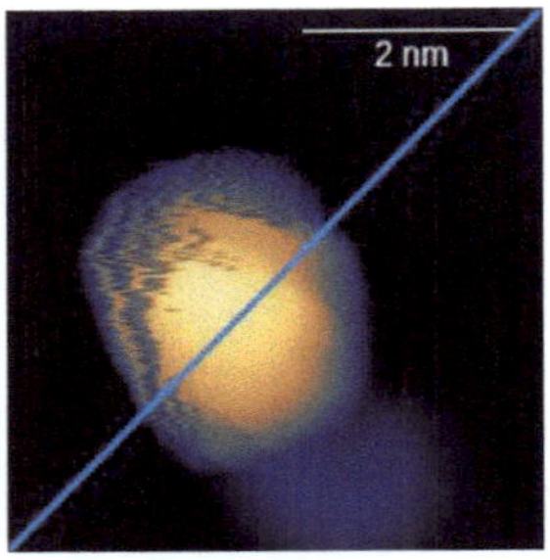

(a) Single sTpd-L at +2 V, I_{set} = 40 pA.

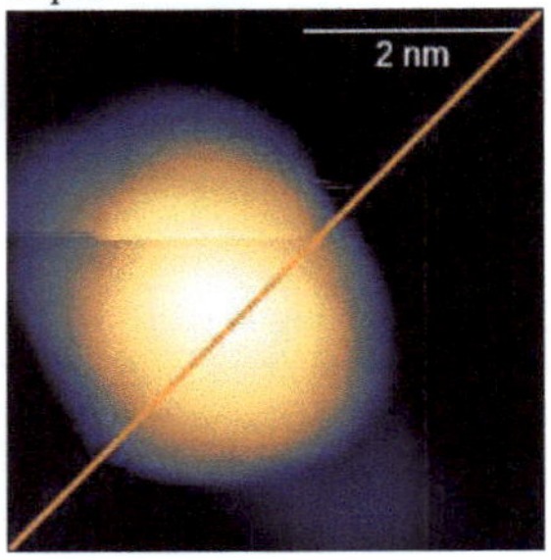

(b) Single sTpd-L at −2 V, I_{set} = 40 pA.

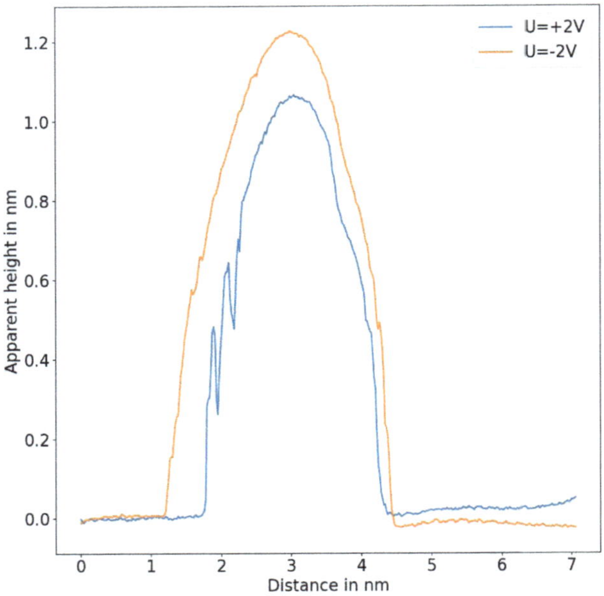

(c) Line profiles indicated in 6.6a and 6.6b. The profile at negative bias is wider, suggesting that the polar head group is pointing away from the surface.

Figure 6.6: Comparison of widths of single sTpd-L molecules.

In conclusion, sTpd-L adsorbs with the functional group on the surface in islands, because the interaction between multiple head groups stabilizes a lying configuration. However, single sTpd-L molecules are standing upright. Therefore, it seems to be at the verge of being able to carry a functional group comparable to the chiral Geländer head group. This suggests, that another version of tripodal platform has to be chosen to support the proposed chiral Geländer head group.

6.2 Extended Tetraphenylmethane-based Platform Tpdx

This section presents results on molecules with an extended tripodal (Tpdx) platform with various head groups. Combining the information gathered from these studies, a combined deposition of two types of molecules provides samples for which the functional group points away from the surface. This also yields a much higher probability of observing directed rotation of the head group.

6.2.1 Tpdx-H: Platform Without Functional Group

The simplest head group that can be chosen is a single hydrogen atom. Such a molecule is rather flat and allows to study the Tpdx platform itself in more detail. More specifically, the arrangement in ordered islands can be studied, providing a reference for further studies on molecules with other functional groups. Intuitively, the arrangement in ordered molecular islands is expected to change drastically, when the functional group lies on the surface and is therefore incorporated in the lattice. Thus, if the arrangement remains the same for molecules with different functional groups, it is a good indication that the adsorption of the three legs with the anchor groups has not changed. More specifically, it indicates that the Tpdx platforms form the lattice and the functional group is pointing away from the surface.
The molecular structure of Tpdx-H is depicted in Fig. 6.7a. The single legs of the Tpdx platform not only consist of a benzene ring with a SAc anchoring group at the meta position (as for the Tpd platform), but are elongated by an extra benzene ring connected with two triple-bonded carbon atoms, resulting in a tolane-derivated structure. The larger footprint of this molecule makes it more likely to support the chiral Geländer head group.
Upon deposition onto the Au(111) surface, Tpdx-H forms ordered islands. A large-scale image is shown in Fig. 6.7b. Again, contaminants from the deposition process are visible. Moreover, comparably high objects can be found, which are most probably fragments of Tpdx-H lying on top of the ordered structure.
As shown in Fig. 6.7b, the ordered structures arrange in domains of different orientations. The orientations visible here are along the $(11\overline{2})$-direction of the Au(111) surface ($< 1\,^\circ$ deviation) and the other orientations are at approximate angles of $30\,^\circ$

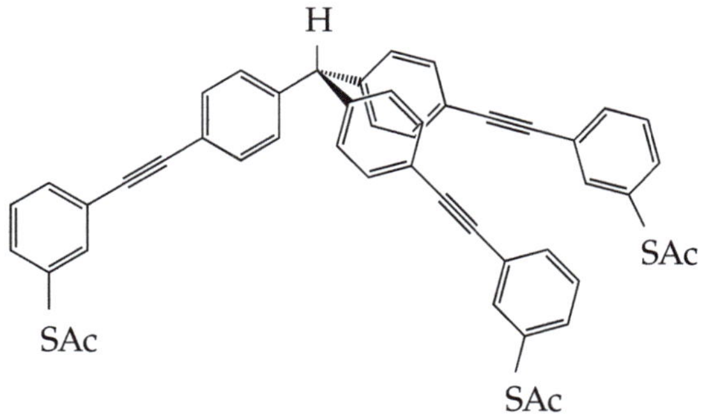

(a) Chemical Structure of Tpdx-H. The flat structure allows for studies of the arrangement in ordered islands for later reference.

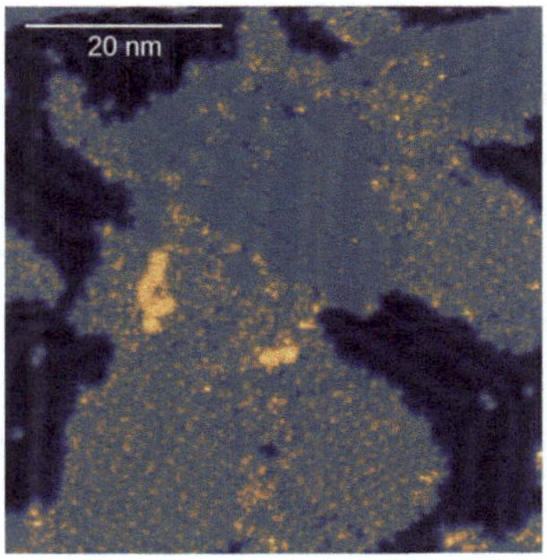

(b) Overview of Tpdx-H on Au(111). Different orientations of ordered islands are present ($U = 1.5$ V, $I_{set} = 50$ pA).

Figure 6.7: Chemical structure and overview of Tpdx-H on Au(111)

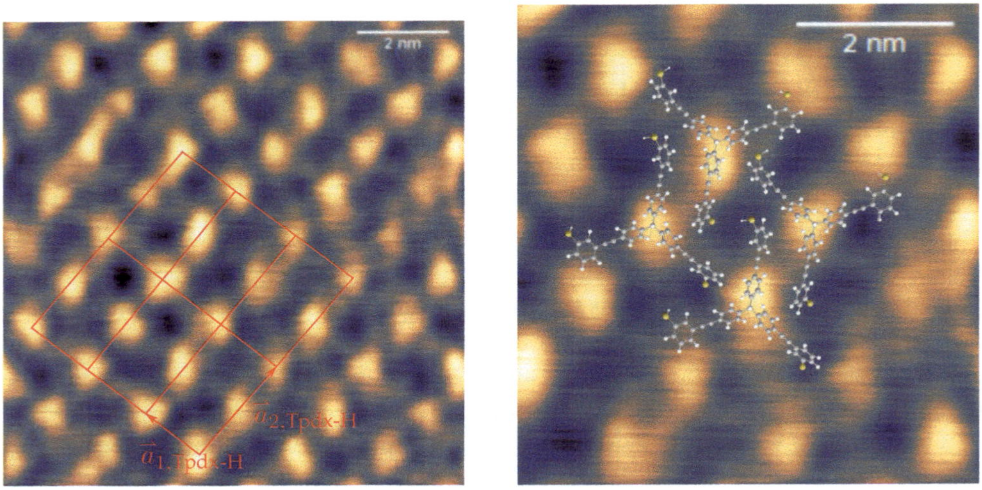

(a) Close-up of ordered island of Tpdx-H with indicated unit cells and lattice vectors.

(b) Same image as 6.8a with molecular structure superimposed to scale.

Figure 6.8: Ordered islands of Tpdx-H (both at $U = -0.5\,\text{V}$, $I_{\text{set}} = 100\,\text{pA}$).

and 90 °. This indicates, that the islands are not always aligned along a symmetry axis of the Au(111) surface and six different orientations with respect to the surface are expected.

Taking a closer look at the lattice formed by Tpdx-H (see Fig. 6.8a), it seems to be rather rectangular than hexagonal. At first sight, this is unlikely, as both the surface and the molecule have three-fold symmetry. However, superimposing the molecules to scale onto the lattice as in Fig. 6.8b, the arrangement matches the lattice very well. Also, the substructure which is visible for individual molecules within the island matches the extent of the legs of the superimposed molecules reasonably well. Here, the Ac groups were omitted in the superimposed molecules for clarity.

Indicated in Fig. 6.8a is also the unit cell of the ordered structure. The lattice vectors have lengths of $|\vec{a}_{1,\text{Tpdx-H}}| = 1.55\,\text{nm}$ and $|\vec{a}_{2,\text{Tpdx-H}}| = 2.5\,\text{nm}$ enclosing an angle of $89.5\,°$. The unit cell's size is $3.87\,\text{nm}^2$ and it contains two molecules, resulting in a molecular footprint of $1.94\,\text{nm}^2$. Comparing this footprint to previous molecules [82], it is much larger. This makes it more likely for the Tpdx platform to be able to support the Geländer head group.

6.2.2 Tpdx-S: Symmetric Functional Group

In a next step towards investigating the rotation of Geländer head groups of single molecules, a symmetric molecule was studied. In this case, symmetric refers to the head group being the same as the three foot structures. Finding characteristic properties of these molecules could help to distinguish molecules adsorbed with all three feet on the surface from those adsorbed with the head group on the surface (called 'standing upright' and 'lying down', respectively, in the following). The corresponding chemical structure is depicted in Fig. 6.9a.

As shown in Fig. 6.9b, Tpdx-S adsorbs both in ordered islands and as single molecules, both exhibiting comparably large apparent heights of $\sim$ 1 nm. This compares well to single sTpd-L molecules, which is plausible due to the similar lengths expected from the chemical structures. Moreover, it indicates that the Tpdx-S molecules observed are intact.

The slightly lower apparent height may arise from the SAc group at the end of the functional group. It has no dipole moment and therefore no electrostatic interaction influencing the apparent height. Also, if the Ac is cleaved off during preparation, the remaining group is, at least from the chemical structure, shorter than the cyanide group of sTpd-L. Unfortunately, previous experiments where the Ac group in para-position could be cleaved off *via* voltage pulses and S–S bonds between molecules formed [82], could not be reproduced with these molecules. Most likely, the length of the functional group and the rotational freedom of the SAc group in meta-position prevent the cleavage and the S–S bond formation.

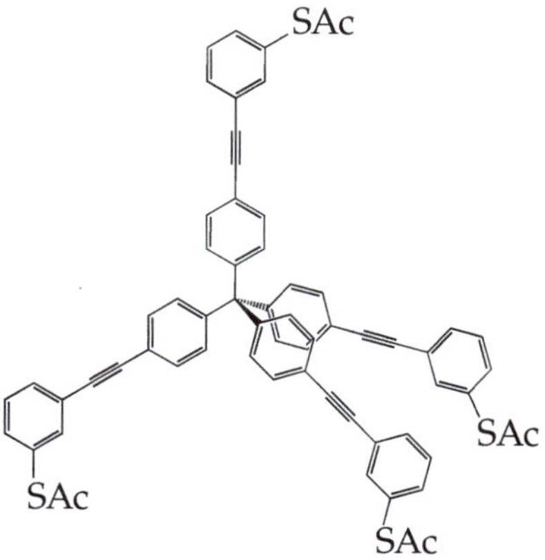

(a) Chemical structure of Tpdx-S.

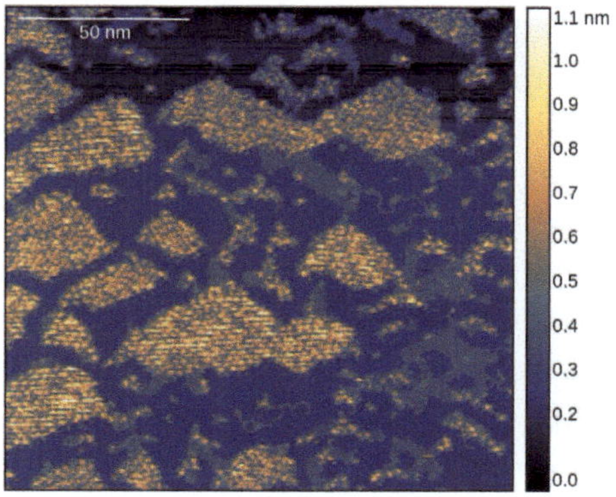

(b) Tpdx-S on Au(111). The ordered islands show different orientations (U = 2 V, I_{set} = 10 pA).

Figure 6.9: Chemical structure and overview of Tpdx-S on Au(111).

As for Tpdx-H, the islands of Tpdx-S form domains of different orientations. Extracted from Fig. 6.9b, the three observed orientations enclose angles of $28\,°$, $98\,°$ and $151\,°$ with the $(11\bar{2})$-direction of the Au(111) surface, which is compatible with the possible orientation directions of Tpdx-H.
A more detailed image of the order of Tpdx-S can be found in Fig. 6.10a. Here, also the proposed unit cell is indicated, together with the corresponding unit vectors. They have lengths of $|\vec{a}_{1,\text{Tpdx-S}}| = 1.61\,\text{nm}$ and $|\vec{a}_{2,\text{Tpdx-S}}| = 2.50\,\text{nm}$, enclosing an angle of $91.1\,°$. The unit cell's size is $4.02\,\text{nm}^2$ and it contains two molecules, resulting in a molecular footprint of $2.01\,\text{nm}^2$. These values are within a reasonable error margin compared to the ones observed for Tpdx-H, emphasizing that even with a large head group mounted on the platform, the order in the formed islands remains the same.
A difference in the ordered structures of Tpdx-H and Tpdx-S is a shift between the two blobs representing Tpdx-S molecules within the unit cell, which is not present for Tpdx-H. It is most probably due to different tilts in the rather long head group combined with steric hindrance effects. Previous studies [85] and unpublished DFT calculations from the group of Prof. F. Evers (see Sec. 7.5) indicate that the head groups are most probably in a configuration where they are not exactly perpendicular to the surface. Also, the molecular structure allows for a flexibility in the exact position of the head group, so that it can, for example, have metastable configurations in between or on the feet of the platform. Therefore, if many head groups are arranged close to each other, they interact with each other, resulting in slightly shifted positions of the apex of the head group, which mainly defines the position of the blob visible in STM images.
In both the ordered islands and single molecules, Tpdx-S appears more like a roughly circular blob and shows no recognizable substructure. The structure is most probably a superposition of metastable states of the functional group. They switch frequently during scanning, being pushed from one state to the other by the tip, which results in numerous sharp edges in the STM images, as seen in Fig. 6.10. Furthermore, the visible shape is a convolution of the molecular structure with the tip shape. This results in non-reproducible shapes in the images, rendering it extremely difficult to find characteristics to distinguish these molecules from the images alone. Also, other attempts to find characteristics of the Tpdx-S molecule, such as differential conductance spectra or constant height images did not yield reliable results.
In conclusion, no characteristics of a foot structure pointing towards the tip could be identified. However, it could be observed, that Tpdx-S forms islands very similar to the ones observed in Tpdx-H. This shows, that head groups as large as a foot group can still be incorporated in ordered islands without disturbing the order.

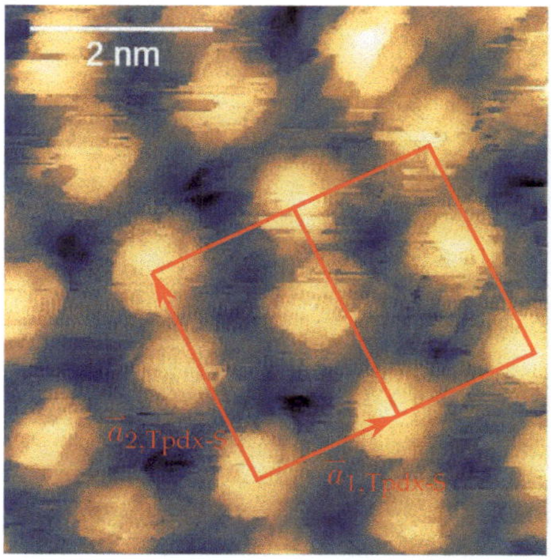

(a) Tpdx-S ordered structure with unit cell.

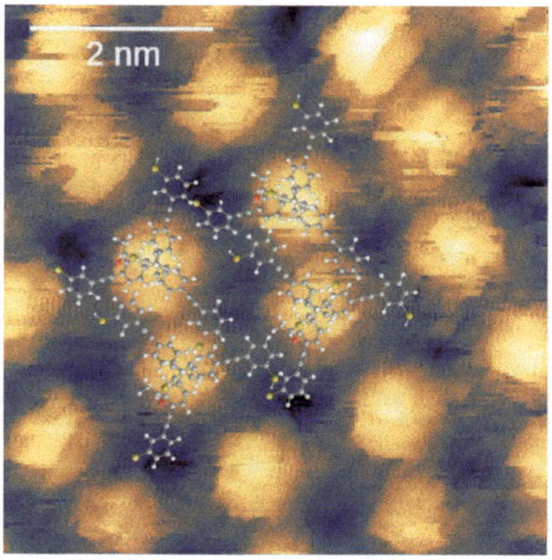

(b) Tpdx-S as in Fig. 6.10a with superimposed molecular structure to scale.

Figure 6.10: Ordered islands of Tpdx-S on Au(111) (both at $U = 1.5\,\mathrm{V}$, $I_{\mathrm{set}} = 6\,\mathrm{pA}$).

6.2.3 Tpdx-G: Geländer Functional Group

With the promising results on how Tpdx-H and Tpdx-S behave when deposited on the Au(111) surface, the attempt to study a molecule, where the Geländer head group is mounted on the Tpdx platform, is sensible. The corresponding chemical structure is depicted in Fig. 6.11a. If the Geländer head group points away from the surface, Tpdx-G is suitable to study the rotation of the functional group.
As for the previous molecules, an overview of Tpdx-G on Au(111) is shown in Fig. 6.11b. Again, both ordered islands and single molecules are observed. For Tpdx-G there are again more than three orientations of the ordered islands. This suggests that, since Tpdx-H and Tpdx-G have more than three orientations, other orientations were not observed for Tpdx-S but were most probably also present.
Comparing the lattice parameters of the ordered islands to the ones observed in Tpdx-H and Tpdx-S, shows that they are again in good agreement. A close-up of an ordered island with the proposed unit cell and the same image with superimposed molecules can be found in 6.12a. The lattice vectors have lengths of $|\vec{a}_{1,\mathrm{Tpdx\text{-}G}}| = 1.57\,\mathrm{nm}$ and $|\vec{a}_{2,\mathrm{Tpdx\text{-}G}}| = 2.52\,\mathrm{nm}$, enclosing an angle of $88.7\,°$. The unit cell's size is $3.96\,\mathrm{nm}^2$ and it contains two molecules, yielding a molecular footprint of $1.98\,\mathrm{nm}^2$, in full agreement with the lattice parameters of both Tpdx-H and Tpdx-S. As mentioned before, such similar values for the lattice parameters strongly suggest, that the ordered lattice is formed by the Tpdx platform for all three molecules. This implies that the Geländer group points away from the surface in the ordered islands of Tpdx-G. A superposition of the molecule onto the lattice is shown in Fig. 6.12b.

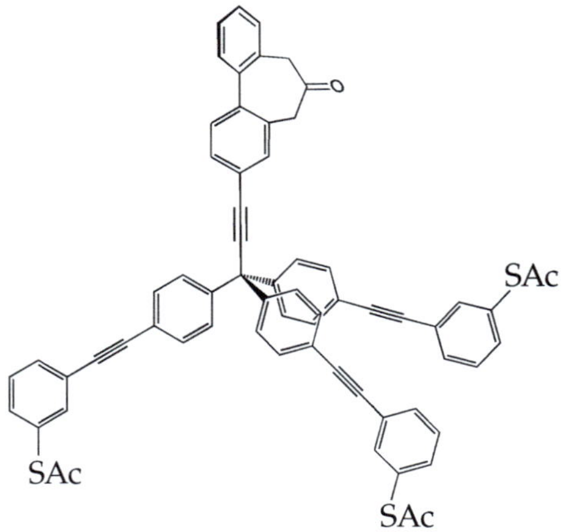

(a) Chemical structure of Tpdx-G.

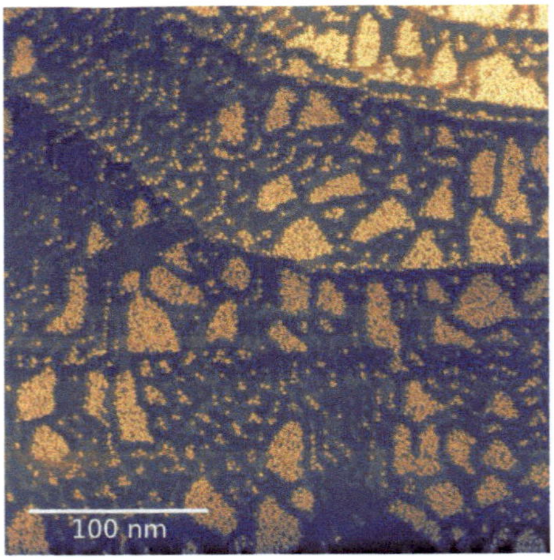

(b) Overview of ordered islands and single molecules of Tpdx-G (U = 1.5 V, I_{set} = 30 pA).

Figure 6.11: Chemical structure and overview of Tpdx-G on Au(111).

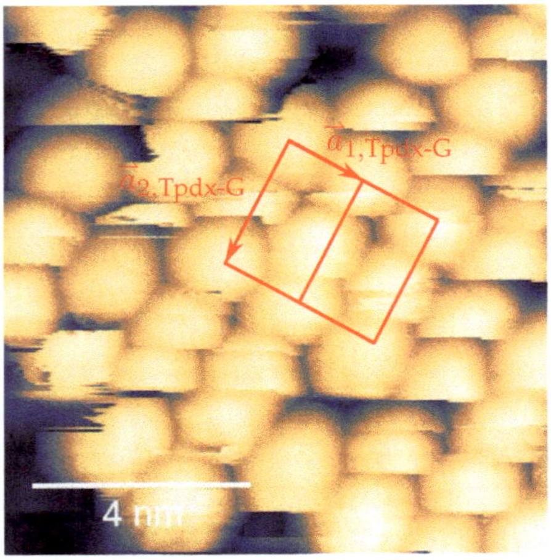

(a) Ordered island of Tpdx-G with indicated lattice vectors and corresponding unit cells.

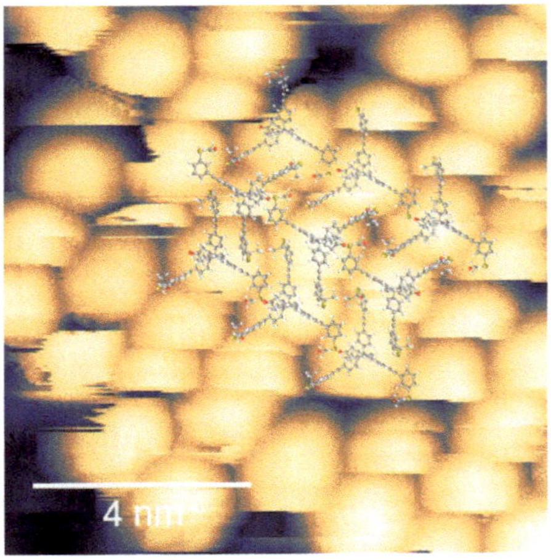

(b) Same ordered islands as in 6.12a with molecular structure superimposed to scale.

Figure 6.12: Ordered islands of Tpdx-G on Au(111) (both at U = 1.5 V, I_{set} = 10 pA).

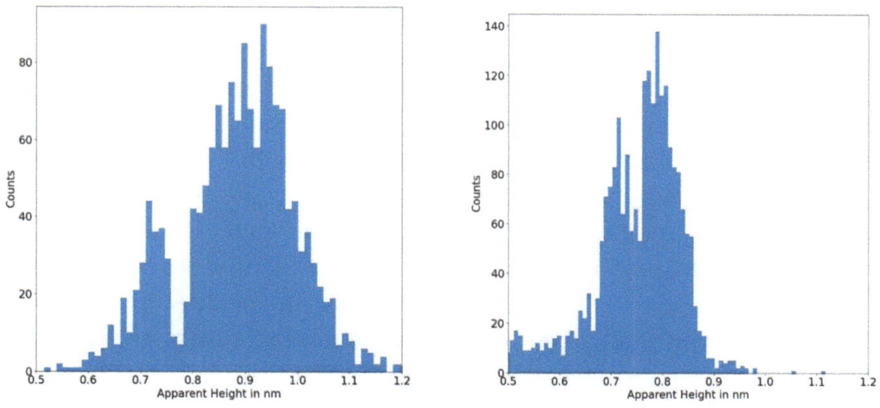

(a) Tpdx-S. (b) Tpdx-G.

Figure 6.13: Distribution of observed apparent heights for Tpdx-S and Tpdx-G. The distributions are too wide and not very well separatable to be able to judge whether a foot group or the Geländer group is pointing away from the surface.

From Figures 6.11b and 6.12 it becomes apparent, that the ordered structure of Tpdx-G has more defects in the lattice than the ones of Tpdx-H and Tpdx-S. This might be a result of the more rigid head group combined with steric hindrance effects that make it energetically unfavorable to have the Geländer groups arranged parallel to each other and in close vicinity. Also the partial charge caused by the electronegativity of oxygen in the functional group might contribute by causing repulsive forces.

Even though Tpdx-G is most likely to adsorb with the platform on the surface within ordered islands, this conclusion cannot be transferred to single molecules. As the apparent height of Tpdx-G ($\sim 0.8\,\mathrm{nm}$) is similar to Tpdx-S ($\sim 0.9\,\mathrm{nm}$) at the same tunneling conditions and they have too wide height distributions (see Fig. 6.13), it cannot be used as a characteristic to distinguish between standing and lying molecules. Also, as mentioned in Subsec. 6.2.2, no other characteristics of molecules with one of the three anchoring groups pointing away from the surface could be identified.

To be able to study the rotation of the head group, however, it is essential to study single molecules. This has multiple reasons. If embedded in an ordered island, the functional groups of the adjacent molecules will interact with each other, interfering with each others rotation. Intuitively, this would lead to a hindrance of the movement of the functional groups. Moreover, the current injected by the tip could partially run through adjacent molecules, distorting the relationship between current and movement of the molecule.

For the case that the Geländer head group is pointing away from the surface, it should be possible to check, whether rotational behavior can be observed. Indeed, in rare cases, abrupt changes between three or more current levels can be observed when keeping the tip at a fixed position above single Tpdx-G molecules. The three levels are interpreted as metastable states of a full 360 ° rotation of the head group (see Sec. 7.1). Measuring the current continuously at a fixed tip position for $10 - 30$ min is called time trace (TT) measurement in the following. From these TTs, it is possible to judge, whether the rotation is directed and to extract characteristics of the rotation. In one example (see Sec. 7.1) the number of switches in one direction and the other are 117 and 298, respectively. Even without accurate statistical analysis, it is rather clear that one direction is preferred over the other.
As already pointed out, statistical analysis of the rotation is necessary. This requires a large database, to draw conclusions from the data. However, during the measurements on single Tpdx-G molecules, only few TT measurements could be performed. This can have various reasons, including interaction with neighboring molecules, proximity to impurities on the surface or it could indicate, that single molecules can still adsorb with the functional group on the surface. If only few molecules allow for TT measurements, this causes the problem that large amounts of time have to be spent to acquire enough data to draw relevant conclusions.
On another note, the switches between different current levels in TT measurements take place on a time scale of seconds, which is too fast to image the head group in one state. Also, no tunneling parameters could be found to avoid switching during a scan. Therefore, Tpdx-G does not allow for imaging of single metastable states and the direction of rotation cannot be identified as clockwise or counterclockwise.
However, the molecule itself seems to be a promising candidate to investigate the rotation of a chiral functional group. To do this in a straightforward way, it would be beneficial to change the molecular structure such, that single states can be imaged, as attempted in the studies of Tpdx-Gx presented in Subsec. 6.2.4. Another way is to embed single Tpdx-G molecules in a lattice of Tpdx-H molecules to enable the separation of the functional groups and simultaneously ensuring that they adsorb in an upright configuration (see Subsec. 6.2.5).

6.2.4 Tpdx-Gx: Extended Geländer Functional Group

As mentioned in Subsec. 6.2.3, it is not possible to image the single metastable states that the Geländer functional group assumes during TT measurements on Tpdx-G. Trying to image these states, a different molecular design is chosen. The Geländer group is extended by an extra benzene ring at the Geländer head group, therefore the name extended Geländer (Gx). The corresponding molecular structure is shown in Fig. 6.14a.

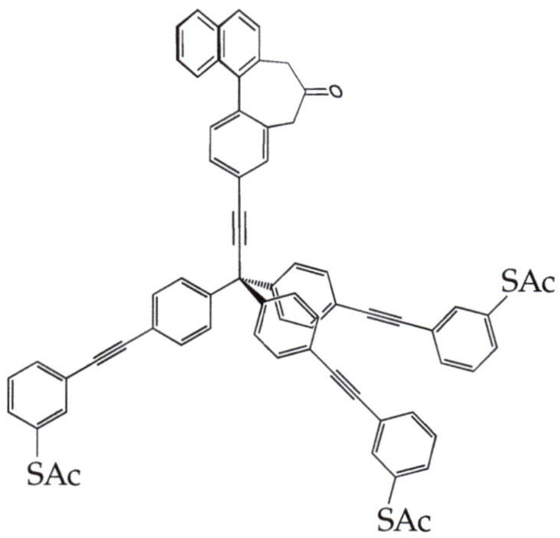

(**a**) Chemical Structure of Tpdx-Gx. The added benzene ring could make imaging of single metastable states possible.

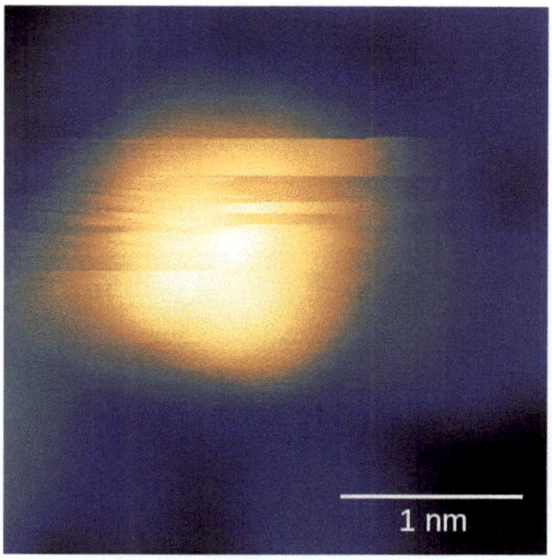

(**b**) Image of a single Tpdx-Gx molecule. It has no clearly asymmetric features, one indication that Gx most likely adsorbs on the surface ($U = 1.5$ V, $I_{set} = 30$ pA).

Figure 6.14: Chemical Structure of Tpdx-Gx and a single molecule on Au(111).

This structure is chosen for multiple reasons. Firstly, as the tip mostly images the closest part of the molecule (especially for objects with high aspect ratios), the added benzene ring should be visible the best at the point of the molecule, which is closest to the tip. This should give rise to an asymmetric structure observable for every molecule, making it easier to distinguish different rotational states. Also, it would allow for a distinction between molecules in lying down and upright configuration. Secondly, the increased size of the functional group should increase the energetic barriers for the switching between different metastable states. Effectively, this would reduce the switching rates between rotational states of the functional group, possibly making single states accessible in imaging.

Lastly, the added benzene ring is not a drastic change to the structure of the molecule, so that the Tpdx platform is hoped to be able to support the extended head group.

Unfortunately, Tpdx-Gx's behavior does not meet the expectations when adsorbed on the Au(111) surface. In contrast to the other Tpdx molecules, no ordered islands could be observed for Tpdx-Gx. Figure 6.15 shows images of Tpdx-Gx on Au(111) at different molecule coverage, ranging from sub-monolayer to possibly multilayer. In none of the images, more than a few molecules are arranged in an ordered fashion. Even at very high coverage, where it should be more probable to find such structures, only disordered agglomerations or layers of molecules form.

For adsorption with the Tpdx platform, ordered islands with similar lattice parameters as for previous molecules (Tpdx-H or Tpdx-S) should be observable. The absence of such ordered islands indicates that the molecules are not adsorbed with

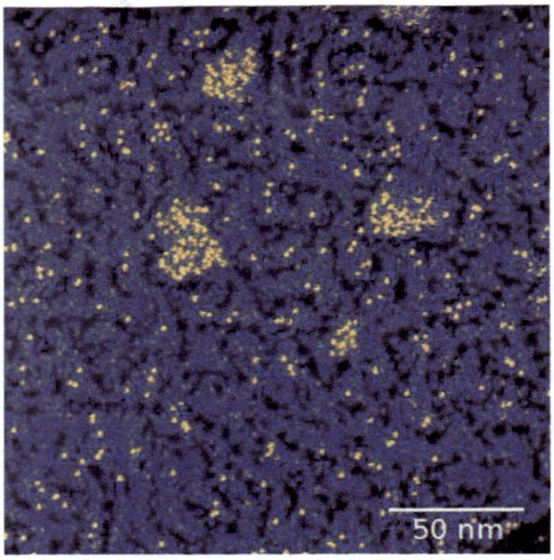

(a) Coverage comparable to previous overview scans.

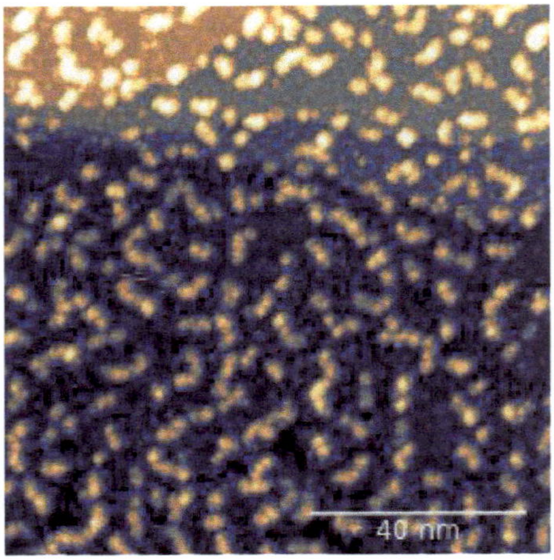

(b) Intermediate coverage.

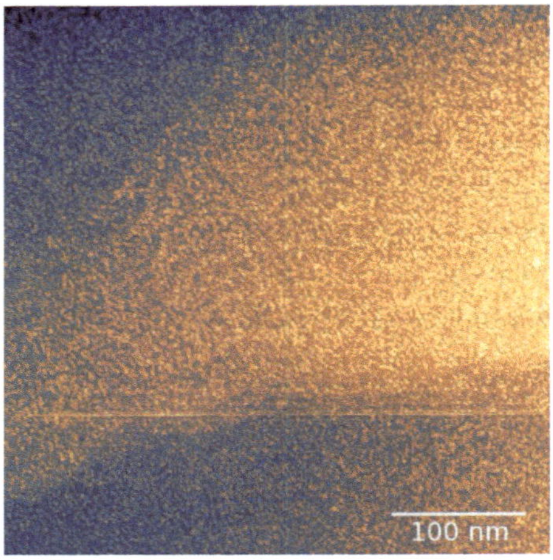

(c) Very high coverage.

Figure 6.15: Unordered islands of Tpdx-G on Au(111) at varying coverage. The absence of ordered islands indicates that the molecules adsorb with the Gx head group on the surface.

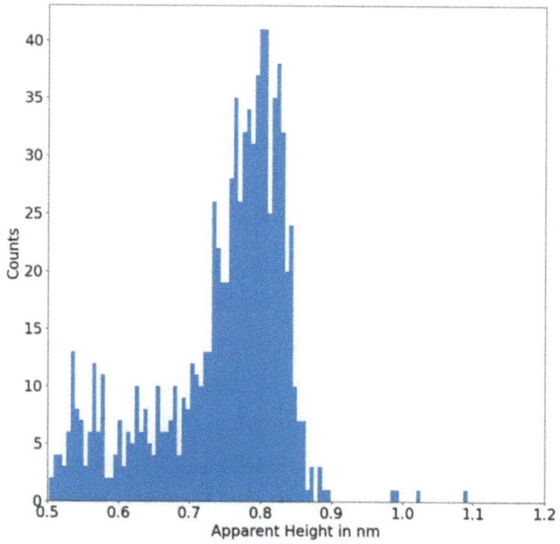

Figure 6.16: Apparent height distribution for Tpdx-Gx for the same tunneling conditions as in Fig. 6.13. The height is distributed around 0.8 nm, which is closer to Tpdx-G than Tpdx-S.

the Tpdx platform, but rather with the Gx head group on the Au(111) surface. This suggests, that it is not the energetically most favorable arrangement to have the Gx group point away from the surface. One explanation could be the size of the Gx head group. The additional benzene ring increases steric hindrance effects, making a parallel alignment of head groups (as it would occur in ordered islands) unfavorable. Also, it increases the van-der-Waals interaction of the head group with the surface. The increased size, together with the partial charge on the oxygen atom on the linker could cause the lying configuration to be more favorable.
Another indication that the Gx head group is adsorbed on the surface is, that there are no reoccurring, asymmetric features visible in single molecules. An exemplary image of a single Tpdx-G molecule is shown in Fig. 6.14b. The observed structures appear very similar to the ones observed in Tpdx-S and Tpdx-G. It does not show any visible, stronger asymmetry, neither can single metastable states be imaged.
Only the observed apparent height contradicts that the molecules adsorb with the head group on the surface. The apparent height of the Tpdx-Gx molecules is distributed around $\sim 0.8\,\mathrm{nm}$ (see Fig. 6.16, same tunneling parameters as for Tpdx-S and Tpdx-G) and not around $\sim 0.9\,\mathrm{nm}$, which is closer to Tpdx-G than to Tpdx-S. Although it cannot be stated with certainty, these observations suggest that Tpdx-Gx's functional group adsorbs on the surface. It emphasizes the balance in the molecular design, that has to be maintained between keeping the functional group pointing away from the surface and the distinguishability of metastable rotational states. As there is no possibility to reliably determine how Tpdx-Gx adsorbs, it cannot be used to study the rotation of the head group.

6.2.5 Tpdx-H/Tpdx-G: Codeposition

As discussed in Subsec. 6.2.1 and 6.2.3, Tpdx-H and Tpdx-G form ordered islands with very similar lattice parameters, urging the question, if they can form a lattice incorporating both molecules. Figure 6.17b shows an image of a deposition of a 3:1 mixture of Tpdx-H and Tpdx-G (called codeposition in the following), forming the proposed intermixed lattice, as expected. Both the flatter Tpdx-H molecules and the comparably high Tpdx-G molecules are incorporated in one ordered island. Such an intermixed lattice has several advantages over the previously described molecules. First and foremost, it can be ensured, that the Geländer group of Tpdx-G points away from the surface. The Tpdx-H lattice incorporates Tpdx-G molecules without disturbing the order significantly. Due to its different chemical structure, the Geländer group would interact very differently with the feet of the Tpdx platform, if it were adsorbed on the surface, breaking the order of Tpdx-H or at least altering its lattice parameters.
However, for the intermixed islands, the unit cell vectors have lengths of $|\vec{a}_{1,\text{codep}}| = 1.52\,\text{nm}$ and $|\vec{a}_{2,\text{codep}}| = 2.50\,\text{nm}$, enclosing an angle of $91.4\,°$. Therefore, the unit cell's size is $3.81\,\text{nm}^2$. It contains two molecules, resulting in a molecular footprint of $1.91\,\text{nm}^2$. These values are very close to the ones of Tpdx-H, validating that its lattice remains mostly undisturbed by the embedded Tpdx-G molecules.
Beyond that, the unit cells for all Tpdx platform molecules that form ordered islands are very similar, strongly suggesting similar adsorption. Since the head group varies between the molecules, the most straightforward explanation is that they adsorb with the Tpdx platform and the functional group points away from the surface.
Another advantage of the intermixed islands is that the Tpdx-G molecules are well separated from each other. Compared to the islands of only Tpdx-G (Subsec. 6.2.3), the interaction between the head groups should be negligible for the intermixed islands. Also, if the tip is positioned above a lone molecule, the current runs mostly through that molecule and can only contain small contributions from other Tpdx-G molecules.
Finally, the intermixed islands reduce the extent to which the platform of Tpdx-G, being embedded in a matrix of Tpdx-H, can change the configuration by reducing the number of degrees of freedom of the platform. As the two benzene rings of the Geländer group are decoupled from the platform via a C-C triple bond, their rotational moveability should not be hindered significantly. The interaction between the platform and the head group is comparable between Tpdx-G molecules, because single Tpdx-G molecules within a Tpdx-H island are surrounded by the same platforms. In other words, the surrounding molecules are roughly in the same configuration for Tpdx-G molecules within the islands, resulting in a comparable potential landscape for the rotation of the chiral head group. Moreover, it prevents

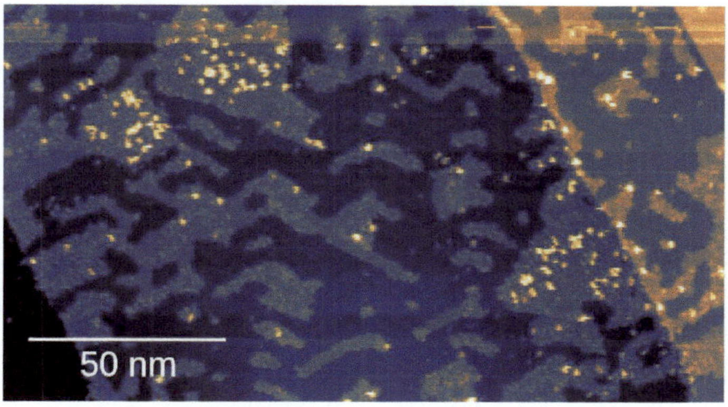

(**a**) Overview of the codeposition on Au(111). One can see the ordered islands of Tpdx-H with the embedded Tpdx-G molecules.

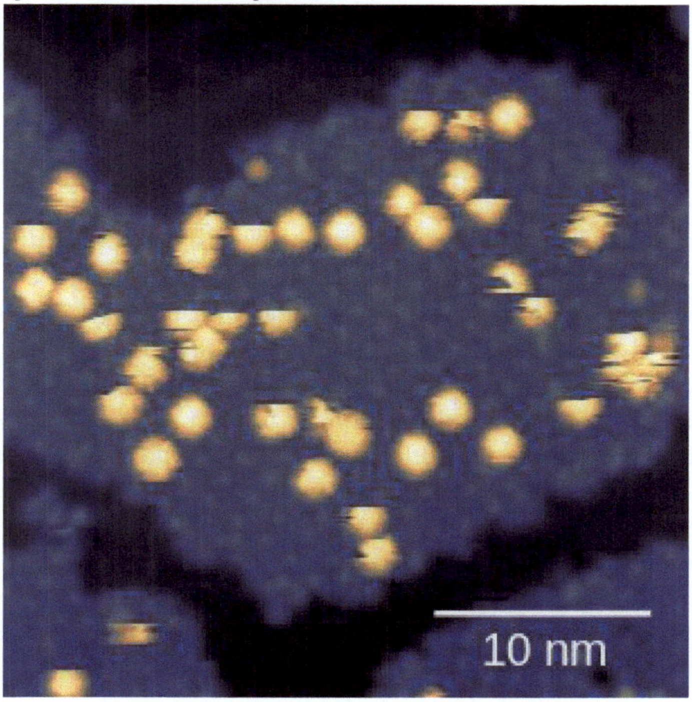

(**b**) Ordered island of the codeposition. The embedded Tpdx-G molecules do not disturb the order of the Tpdx-H island.

Figure 6.17: Structures of the codeposition on the Au(111) surface (both at $U = 1.5\,\mathrm{V}$, $I_{\mathrm{set}} = 30\,\mathrm{pA}$).

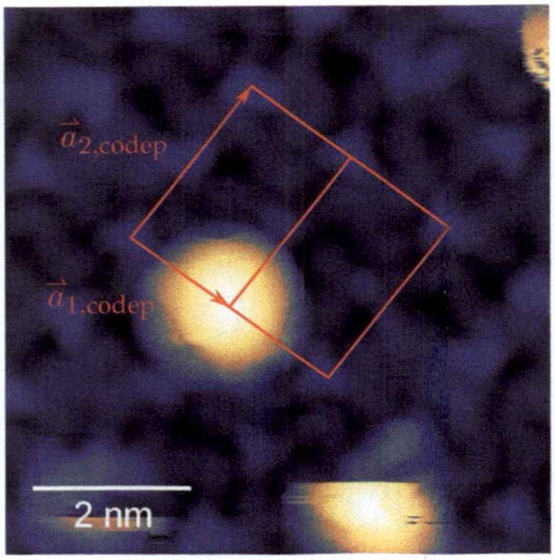

(a) Ordered island of the codeposition. Indicated in red is the unit cell of the Tpdx-H lattice. The center of the Tpdx-G coincides with the lattice.

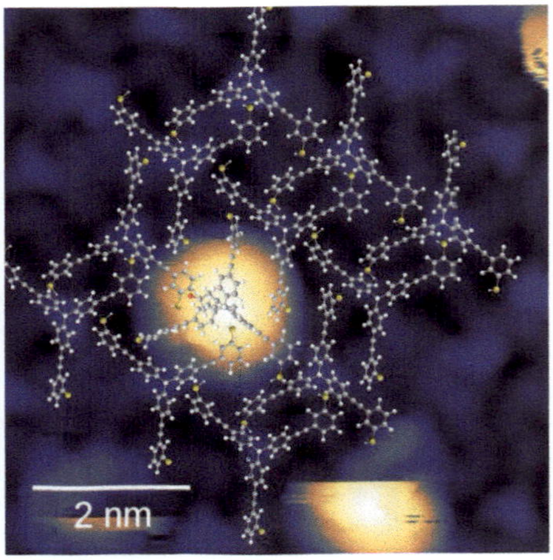

(b) Same image as Fig. 6.18a with both molecules superimposed to scale.

Figure 6.18: Ordered island with unit cell and superimposed molecules (both at $U = 1.5$ V, $I_{\mathrm{set}} = 30$ pA).

Tpdx-G molecules from being surrounded by impurities on the surface. This should reduce the variation in the rotational behavior of the head group, leading to more reproducible and frequent observation of its rotation.
Indeed, as shown in Chap. 7, molecules for which sensible TT measurements could be performed are much more frequent for the codeposition. Out of the measured 425 TT measurements, 180 were suitable for evaluation, which is $\sim 42\ \%$ of the performed measurements. In comparison, for Tpdx-G 114 TTs were measured and only 21 could be evaluated, corresponding to $\sim 18\ \%$. This considerable difference stresses the improved reproducibility for the codeposition. As a result, more types of experiments become feasible, such as TTs for different bias voltages and tip–molecule distances or grid measurements of TTs.
A remaining problem is, that single metastable states can still not be imaged, as expected since Tpdx-G was used. The evaluation still relies on the assumptions discussed in Subsec. 6.2.3.
In conclusion, the combined deposition of Tpdx-H and Tpdx-G, which form intermixed, ordered islands, is very well suited to study the rotation of the chiral Geländer head group of Tpdx-G. The molecules are well separated, avoiding interferences and it is reasonable to assume, that the Geländer head group is pointing away from the surface. Finally, the lattice provides more reliable conditions to study rotational movement. The measurements performed and the evaluation of the results is presented in Chap. 7.

7 Characterization of Rotational Behavior of Tpdx-G

The previous chapter has focused on finding a suitable molecule to study the rotation of a chiral functional group. It is found in the form of a codeposition of Tpdx-H and Tpdx-G, which provides a sufficiently reproducible and stable arrangement of Tpdx-G, so that its chiral group points towards the tip. This way, a current can be run through it and a rotational movement can be excited.

Time trace (TT) measurements, mentioned briefly in Subsec. 6.2.3, recorded on the codeposited Tpdx-G molecules, form the basis of the analysis of the rotational movement. This chapter focuses on the detailed analysis of the rotation. The extraction of the number of rotational switching events from TT measurements and basic evaluation, such as the analysis of the statistical significance are discussed in Sec. 7.2.

Next, other representations of the data, from which characteristics such as the velocity of the movement can be obtained, namely angular trajectories and plots of the mean square displacement, are presented. Scatter plots are examined, which could show correlations between measurement parameters and characteristics of the rotation.

This is followed by the dependence of TT measurements on the bias voltage, on the z position of the tip and on the lateral position of the tip, as well as on the polarity of the bias voltage.

Finally, theoretical calculations of the molecular orbitals of Tpdx-G, performed by Dr. Jan Wilhelm in the group of Prof. F. Evers in Regensburg, and their possible connections to the rotational behavior are discussed.

7.1 Interpretation of Time Trace Measurements

As mentioned in Subsec. 6.2.3, three or more current levels can be observed when keeping the tip at a fixed position above single Tpdx-G molecules. Part of a measurement of the tunneling current taken at such a position is shown in Fig. 7.1.
In this type of measurement, called time trace (TT) in the following, the tip was kept at a fixed position in all three dimensions, the bias voltage remained unchanged and the tunneling current was recorded over time. Based on such measurements, the analysis of the rotational movement is done by interpreting the current levels as metastable states of a full $360\,^{\circ}$ rotation.
If the current remains in one of the three current levels, it means that the functional group remains in the same metastable configuration, which corresponds to one rotational state. A switch between the current levels then corresponds to a change in configuration an thus in a partial rotation of the functional group. This assumption can be justified by looking at the tip–molecule distance during a rotation of the head group.
As indicated in Fig. 7.2, if the tip is at a fixed position and the shown Geländer group changes from one metastable state to another, the tip–molecule distance changes. In turn, this changes the tunneling current due to its exponential dependence on the distance (see Eq. 2.8). As these measurements are done at a fixed tip position and constant bias voltage, the only way to change the tunneling current in the way described before, is by a change of the molecule's configuration. All other changes in current, such as tunneling into another orbital, would have been caused by a change in the bias voltage or tip position.

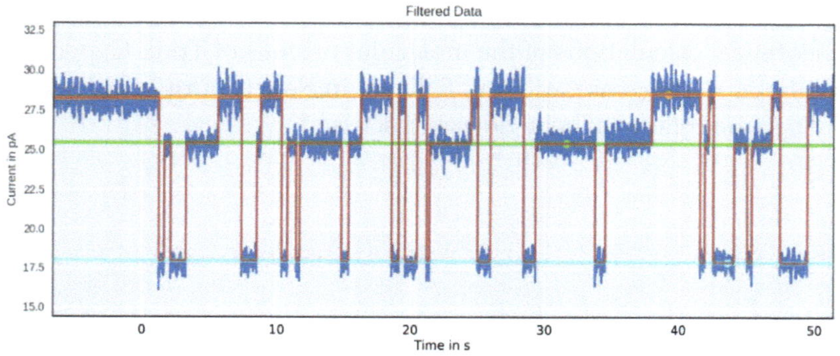

Figure 7.1: Part of a $\sim$ 10 min TT measurement. The filtered signal is depicted in blue, the three defined current levels are cyan, green and orange. The red curve is the idealized measurement.

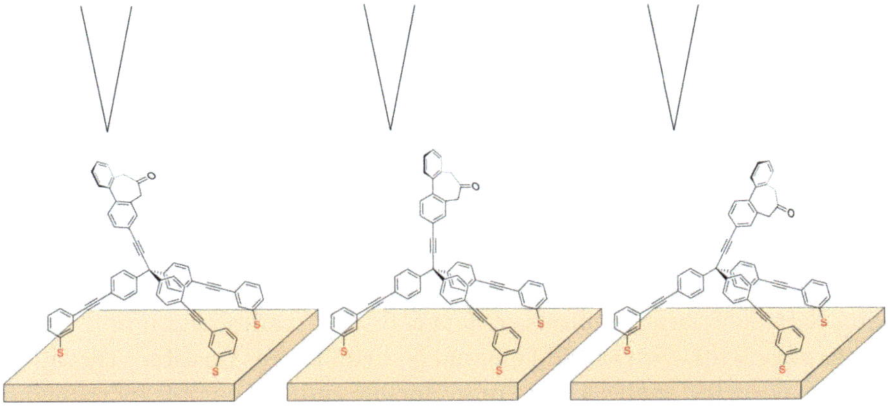

Figure 7.2: Sketch of the rotation of a functional group during a TT measurement. The changing tip-molecule distance due to the different metastable states abruptly changes the current.

Based on this, if three current levels are observed, a switch from one to the other is interpreted as a 120 ° partial rotation of the head group. Then, the switch between the two current levels in the opposite direction would correspond to a -120 ° rotation. Enumerating the three states from 0 to 2, 0 having the lowest current and 2 having the highest, a full rotation would correspond to a switching pattern of $0 \to 1 \to 2 \to 0$ in one direction and $0 \to 2 \to 1 \to 0$ in the other direction.

By counting the switches in one direction and the other and performing a binomial test, it can be determined, whether the surplus of partial rotations in one direction is statistically significant and characteristics of the rotation can be extracted. A detailed description of the full process of data analysis of TTs is discussed in Sec. 7.2, and the figures of merit determined from TT measurements are presented in Sec. 7.3.

Unfortunately, Tpdx-G does not allow for imaging of single metastable states. Therefore, the direction of rotation cannot be identified as clockwise or counterclockwise. It can only be shown that the rotation in one direction is preferred with statistical significance.

Moreover, it cannot be stated that the three configurations of the molecule are the only ones that it can assume. In fact, upon changing the parameters, it was possible to observe four different current levels and hence four metastable states. Still, in the case of three distinguishable states, the rotation can be interpreted as mentioned before.

7.2 Signal Processing

The rotation observed in Tpdx-G is a stochastic process. It is very similar to one-dimensional diffusion with a driving component in one direction. This makes it a necessity to perform statistical analysis. Also, because the molecules react very sensitively to the tip position, there is a large variation in the results upon changing the tip position or measuring on a different molecule. Therefore, large amounts of TT measurements have to be evaluated to see trends e.g. with the bias voltage.
Working with such large amounts of data calls for an automation of the analysis of TT measurements with as little manual contribution as possible. This adds the benefit of making the analysis more reproducible and comparable by reducing the influence of human bias.
However, some parameters of the evaluation have to be adjusted manually for each individual TT measurement. For example, the current levels corresponding to the metastable states are different and can vary over time. As the TT measurements are as long as 10 to 30 min, also thermal drift of the tip–molecule distance has to be compensated. Automating these tasks is challenging and time-consuming, which is why they are still done manually. After the parameters are set, the program does the data analysis automatically.
A solution combining the need for manual settings and automated data analysis is a graphic user interface (GUI). That way, the necessary parameters can be set easily by hand and the program calculates the results reproducibly. A picture of the interface during the evaluation process is shown in Fig. 7.3. It was programmed in python[1] using PyQt5[2] for the graphic interface in the framework of the Master thesis of Philipp Markus [86].
The first step of the analysis is to reduce the noise on the signal. This is done by a fast Fourier transform (FFT) filter. For this purpose, the data is transformed into Fourier space, showing characteristic frequencies in the signal. Sharp peaks are present in the FFT caused by noise at specific frequencies. The user can choose a frequency at which to cut the FFT. The signal above the chosen frequency, once it exceeds a certain FFT amplitude, will be set to zero. The signal is transformed back into the time domain and the filtered signal is retrieved. As the switching between current levels is a stochastic process and does not appear at a fixed frequency and remains mostly untouched by this process.
In the next step, the current levels have to be defined. As mentioned above, the current levels can change over the course of a measurement, which is why they are defined as a polynomial of variable degree. The polynomials are defined by points

[1] `https://www.python.org/`
[2] `https://wiki.python.org/moin/PyQt`

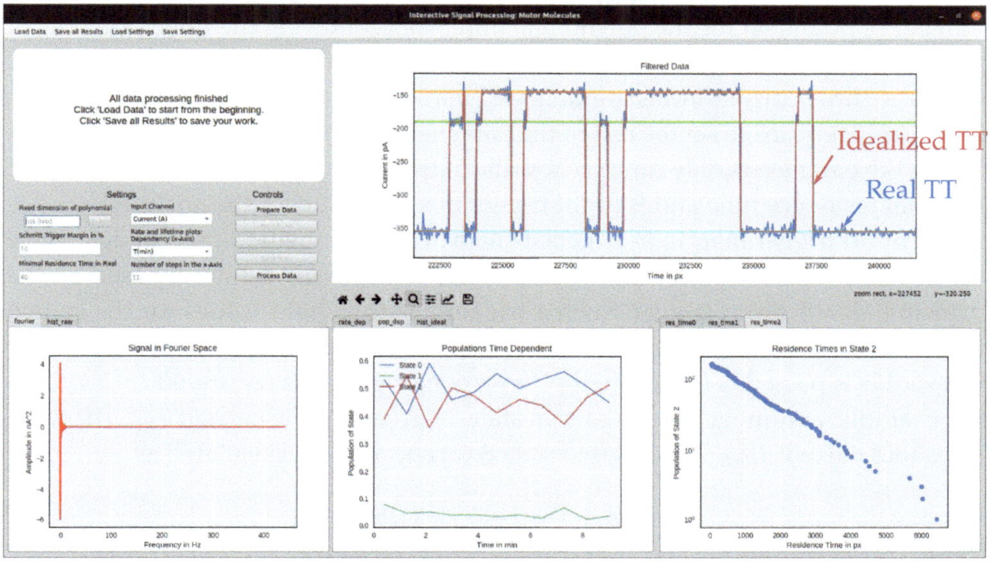

Figure 7.3: The GUI used for the analysis of TT measurements. It shows the signal at the top right, the parameters that have to be set for the evaluation at the top left and the results in the bottom row. The arrows point to the real TT (blue) and idealized TT (red).

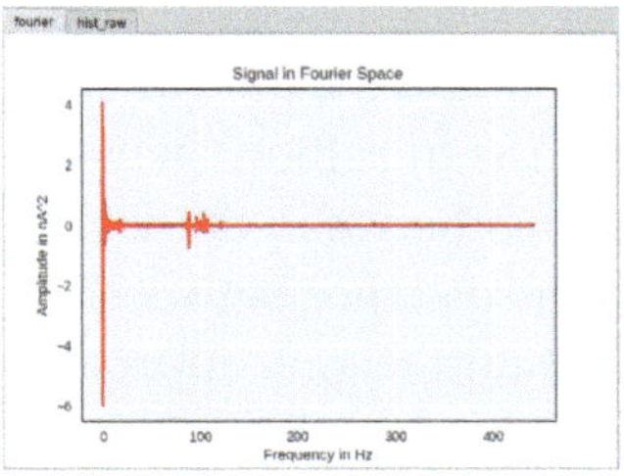

(a) FFT before cutting.

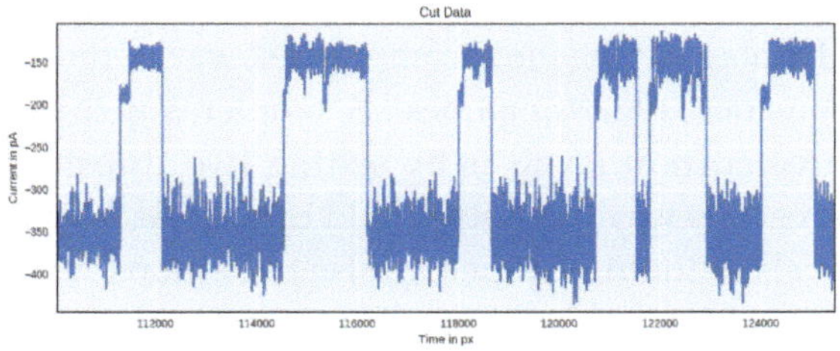

(b) Signal before FFT filtering.

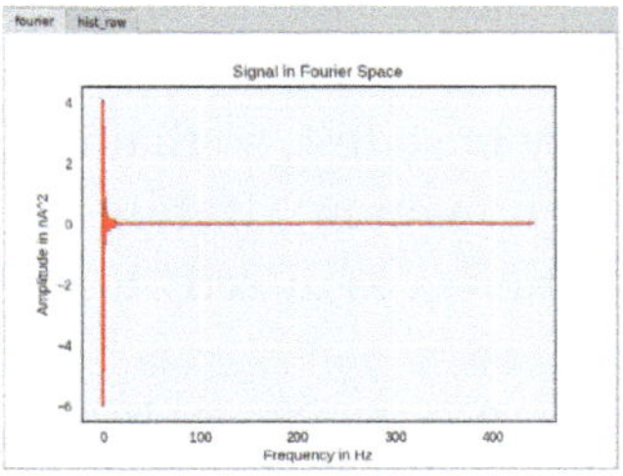

(c) FFT after cutting.

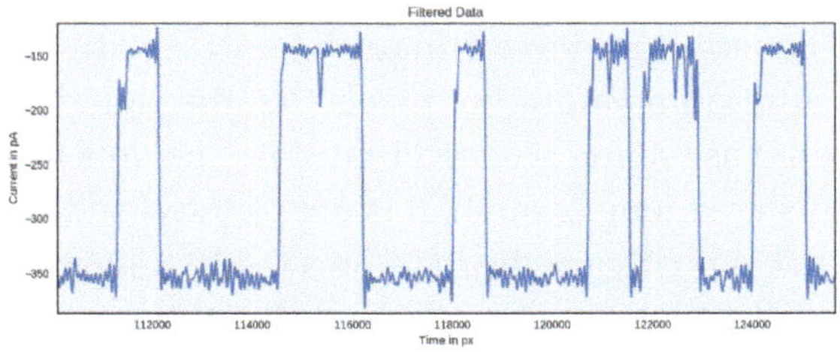

(d) Signal after FFT filtering.

Figure 7.4: FFT filtering before and after cutting the signal in Fourier space.

set by the user by simply clicking into the plot of the signal. The degree is set to be the number of points set for the polynomial minus one, allowing the fitted polynomial to pass through all set points. If desired, the degree can also be fixed to one value. Once the three current levels are defined (marked cyan, green and orange in Fig. 7.3), it only remains to set the rest of the parameters before the analysis can be done. Two of the parameters only concern how the output is shown. However, the so-called minimal residence time and Schmitt trigger margin influence the analysis.
The Schmitt trigger margin m_{ST} gets its name from the method used to determine whether a switch to another current level has taken place, which is based on the principle of a Schmitt trigger. Such a trigger has threshold values for the switch to another level, depending on the level the system is currently in. If one of these thresholds is passed, a change to the corresponding state is registered.
For example, assuming the signal is in state 1 (green) at the current I_{S1}. Then, the threshold current $I_{th,S1\rightarrow S0}$ to go to state 0 (cyan) at I_{S0} is calculated as

$$I_{th,S1\rightarrow S0} = I_{S0} + m_{ST} * (I_{S1} - I_{S0}) \ . \tag{7.1}$$

In other words, the Schmitt trigger margin is the factor to determine the width of current values accepted to be the corresponding state. Therefore, for the transition from state 1 to state 2 (orange) at I_{S2}, the threshold is

$$I_{th,S1\rightarrow S2} = I_{S2} - m_{ST} * (I_{S2} - I_{S1}) \ . \tag{7.2}$$

The remaining thresholds are computed analogously.
The advantage of the more complicated way of finding the points of the transitions is the robustness regarding noise. A simpler way to determine the transitions between the three current levels is by setting two threshold currents between the levels. However, this way the noise could only go as high as half the distance between the two levels without producing false transitions. With the Schmitt trigger approach, the thresholds are state-dependent and the noise can be much higher, depending on the set Schmitt trigger margin.
The other parameter influencing the analysis is the minimal residence time. This time has to be introduced to distinguish $0 \rightarrow 1 \rightarrow 2$ transitions from $0 \rightarrow 2$ transitions. It is the minimum amount of pixels that the signal has to stay in state 1, so that it is registered as having reached that state (i.e. in a $0 \rightarrow 1 \rightarrow 2$ transition). If state 1 is not registered long enough, it is not counted and the transition goes directly to the state that is finally reached (i.e. in a $0 \rightarrow 2$ transition).
If this distinction is not made, no full rotations, such as $0 \rightarrow 2 \rightarrow 1 \rightarrow 0$ can be detected, because the $0 \rightarrow 2$ transition would counted as a $0 \rightarrow 1 \rightarrow 2$ transition. Due to different times spent in state 1 for different TT measurements, the minimal residence time is not the same and it has to be adjusted individually for every measurement. After adjusting these parameters, the transitions can be detected and the idealized

TT (red in Fig. 7.3, scaled to fit currents) is determined. It contains 0,1 or 2 for every point. It is also saved for further analysis steps, such as angular trajectories. Some of the output is mostly used to verify that the measurement is meaningful, such as the so-called residence time plots for each state.

The residence time plots show how often the corresponding state was assumed for a certain time. Because the process is stochastic, the lifetimes should exhibit an exponential decay with time. For clarity, the plots are shown in a semi-logarithmic representation and exhibit linear behavior for a sensible data set (compare Fig. 7.3).

Finally, the numerical results from the analysis are printed to a pop-out window after the processing is done. The most important values are firstly the total amount of transitions, secondly how many of these are so-called ascending transitions, including $0 \rightarrow 1$, $1 \rightarrow 2$ and $2 \rightarrow 0$, thirdly how many are so-called descending transitions, which includes the remaining transitions and lastly the result of the binomial test as a rejection probability (see Subsec. 2.5.1). The binomial test is performed with the values mentioned above and a detailed description can be found in Subsec. 7.3.1. This concludes the analysis done by the GUI.

As a test of the reliability of the analysis done by the GUI, measurements with simulated data were performed. In these measurements, the tip was positioned above the clean gold surface. To simulate the three current levels, the applied bias voltage was switched between three values. The resulting currents ($\sim 60 - 120\,\mathrm{pA}$) and voltages ($1.4 - 1.6\,\mathrm{V}$) were chosen to be close to the ones used in real measurements. The points in time of the switches and the corresponding assumed states are calculated following an exponential distribution (time constants of 5, 6 and 10 px, similar to other TTs) for the lifetimes of the states, as expected for the experiment, and passed to the measurement system (meaning all parts between the application of the bias and the measurement of the current). That way, both the voltage as a reference signal and the measured current can be acquired simultaneously, making them comparable. To ensure that the complete simulated TT does not possess any directionality, the simulated voltage input signal is mirrored in time and appended to itself. From its analysis, one can estimate the error that can be caused by the measurement system and the analysis using the GUI.

Analyzing both the voltage input signal and the current signal with the GUI, the results are quite similar. For the same measurement analyzed with the same settings, the voltage signal gives a total number of transitions of 2981 with 1496 ascending and 1485 descending and the current signal gives 2963 transitions with 1475 ascending and 1488 descending, corresponding to $p_{\mathrm{reject}} = 58.7\,\%$. This means that the current signal not only reproduces the simulated voltage signal well, but also does not introduce a significant false directionality. The same part of the current and voltage signal of the corresponding measurement is shown in Fig. 7.5.

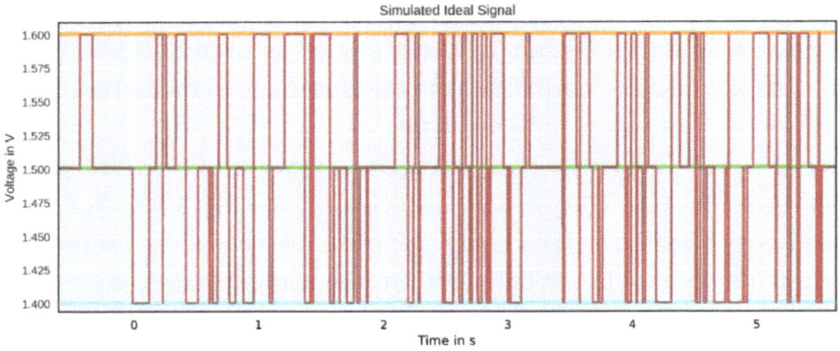

(a) Bias voltage signal of a simulated TT measurement.

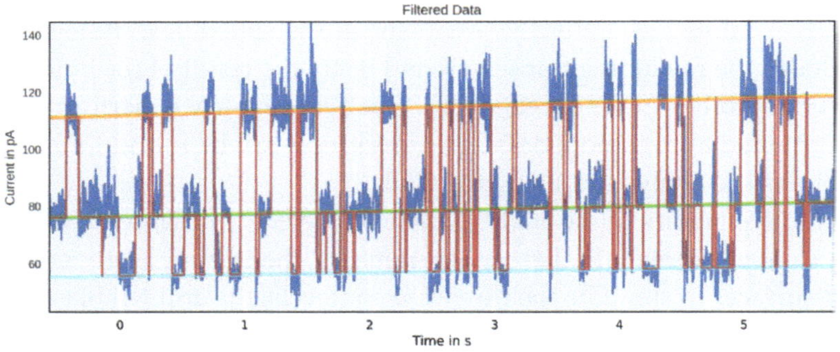

(b) Current signal of a simulated TT measurement.

Figure 7.5: Simulated TT measurement to check the reliability of the analysis.

It becomes apparent from both the numbers and Fig. 7.5, that the evaluation with the GUI represents the original signal reasonably well. Compared to the total amount of transitions of almost 3000, the difference between the voltage and current signal of 18 is small. Also in Fig. 7.5 almost all transitions in the voltage signal are correctly represented in the current signal.

The differences between the signals are caused by residence times in states 0 and 2 that are so short that the averaging of the current trace reduces the current so significantly that it no longer reaches the threshold value to be recognized as a transition. Another cause for errors is the noise introduced by the measurement system which cannot be filtered completely by the FFT filter. This can cause current fluctuations large enough to reach a threshold value to be counted as a transition. Since the signal returns to the original state after such a fluctuation, only symmetric transitions (i.e. $0 \to 1 \to 0$) can result from it. Thus, the latter are not as critical, as they only add to the total amount of transitions but cannot introduce false directionality. Keeping these factors in mind, the analysis using the GUI is reasonably reliable and the influence of the measurement system is small.

To sum up, the GUI that was programmed, using an FFT filter and a Schmitt trigger-like procedure for the evaluation of transitions represents the original signal well and is suitable for the analysis of the rotation of Tpdx-G functional groups. As mentioned before, this makes the results more reliable by removing human bias as much as possible. Moreover, it makes the analysis process faster and more efficient. This tool allows for a fast, basic analysis of the data, but further analysis is needed to extract other meaningful figures of merit of the rotation. These are discussed in the next section.

7.3 Figures of Merit of the Rotation

With the basic evaluation of the data clarified, the next step towards the understanding of the rotational movement of the head groups of Tpdx-G is an assessment of the statistical significance of its directionality. This is discussed in Subsec. 7.3.1, followed by the introduction of another way of representing the data, namely angular trajectories. From these, the velocity of the rotation can be extracted. The theory of diffusive motion offers yet another means of characterizing whether a movement is directional and of finding its velocity, the mean square displacement. It is applied to the rotational movement and its results are compared to those of the angular trajectories in Subsec. 7.3.3.

7.3.1 Binomial Test and Rejection Probability

The goal of this work is to find and study a motor based on a single molecule. The rotational movement of a motor has to be driven in one direction and must not be random. Here, as described in Sec. 2.3, the directional movement is always superimposed on random Brownian motion. The molecule only functions as a motor, if the directional motion is significant compared to the random motion. It is therefore of major importance to be able to distinguish between these types of motion.
According to the interpretation of TT measurements discussed in Subsec. 6.2.3, rotational movement in one direction is represented by ascending transitions and in the opposite direction by descending transitions. The superposition with random motion necessitates statistical analysis. In consequence, the statistical significance of the directionality of the movement can be determined with a binomial test. In this case, the null hypothesis of the test, which is to be proven wrong, is that the observed motion is random.

Translated to what is extracted from TT measurements, this means that the probability to observe an ascending transition would be the same as for a descending transition. For the process described in Subsec. 2.5.1, this corresponds to n being the total amount of transitions n_{tot} observed in the TT measurement, n_+ being the amount of ascending n_{asc} or descending n_{desc} transitions, depending on which is larger, as a one-tailed binomial test is performed, and $p_+ = 0.5$. All the necessary numbers can be taken from the TT evaluation.
The binomial test then provides a p-value and a rejection probability p_{reject} for each TT measurement. Simply put, this can be understood as the probability with which the rotation is directed. p_{reject} ranges from 50%, meaning that rejecting and not rejecting the null hypothesis is just as likely, to approximately 100%, meaning that the null hypothesis can only be rejected. While exactly 100% is unreachable, it is possible to get very close. In the scope of this work, all $p_{\text{reject}} > 99.9999\%$ are approximated to 100%. Such measurements have a significance level of 4.75σ, meaning that such an extreme outcome deviates more than 4.75 standard deviations from the null hypothesis expectation value, which is considered definitive proof here.
With the results from the binomial test, it is possible to get an overview of the frequency of observing a TT with a p_{reject} strongly suggesting directed rotation. Here, a value of $p_{\text{reject}} \geq 99.9\%$ is chosen to be this threshold. It corresponds to approximately '3σ' and is a reasonable value to assume that the null hypothesis can be rejected with statistical significance. Out of the 180 evaluated measurements, 74 had a p_{reject} fitting this criterion. This corresponds to 40% of TT measurements, stressing the reproducibility of them showing directed rotation with statistical significance.
Overall, the binomial test proves a useful tool in evaluating the statistical significance of directed rotation in TT measurements. It can be done easily by counting the types of transitions observed. In the evaluated measurements, a large percentage shows directed motion, strongly suggesting that Tpdx-G shows a preferred direction of motion. For a better quantitative understanding of the motion, it is useful to extract its velocity, which is the topic of Subsec. 7.3.2 and 7.3.3.

7.3.2 Angular Trajectories and Asymmetry

An important characteristic of any kind of rotation is its velocity. However, in this case, it is not straightforward to find a way to define and determine it in a meaningful way. As the rotation is determined by discrete transitions between states, they first have to be translated to a physical quantity. The approach taken here is closely related to the interpretation of the TT measurements, as described in Sec. 7.1.
In this, every transition is interpreted as a rotation of the functional group by $\Delta\varphi = \pm 120\,^\circ \mathrel{\hat{=}} \pm\, 2\pi/3$, depending on the direction of the transition (i.e. ascending or descending). With the discrete transitions translated to an angle, the trajectory of the

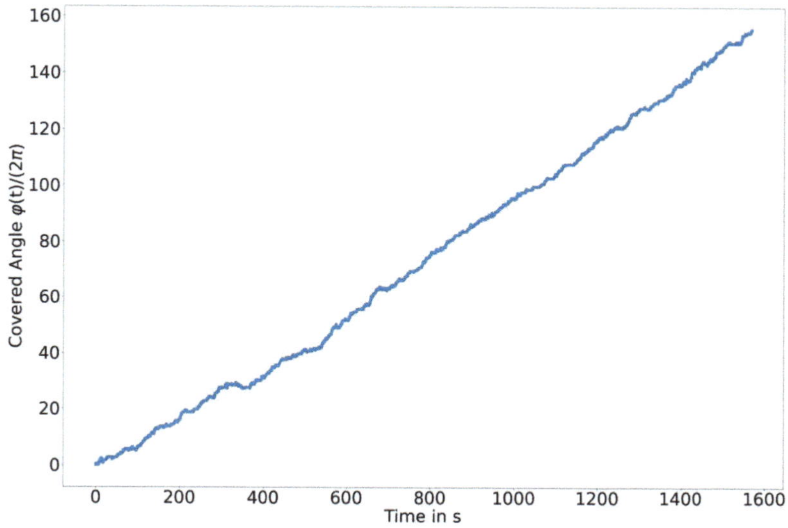

Figure 7.6: Angular trajectory of a TT measurement. Plotted is the covered angle of the functional group in units of 2π as a function of time. The linear behavior indicates driven motion. The velocity of rotation is taken from the slope of a linear fit.

rotation can be calculated. For that purpose, the idealized TT (as mentioned in Sec. 7.2) is used, starting at an angle of zero and adding $\Delta\varphi = \pm 2\pi/3$, depending on the direction of the corresponding transition. By continuously adding or subtracting the angles, one gets φ, the total angle covered by the rotation of the functional group. As the idealized TT also contains the time, the covered angle $\varphi(t)$ can be plotted as a function of time. These plots are called angular trajectories in the following.
An example of an angular trajectory for a rotation with high rejection probability is shown in Fig. 7.6. It is clearly visible, that it exhibits linear behavior. For random, Brownian motion, this plot would follow a square root function, as expected for the average distance travelled from the starting point (compare Subsec. 2.5.2). The linear behavior again supports the conclusion, that the observed motion is driven and therefore has a preferred direction.
Note, that the angular trajectory does not feature inversions of the slope, i.e. going from a positive to a negative slope, meaning that the preferred rotation direction stays the same throughout the complete measurement. Such inversions of the preferred rotation direction, on time scales that cannot be caused by random fluctuations, are not observed for angular trajectories.
As the characteristic property of these trajectories, making them indicative of directed motion, is their linearity, the velocity of rotation ω_{rot} can be extracted as their slope. For that purpose, a linear fit is performed, allowing for the calculation of the velocity.

It is usually given in full rotations per second $\left(\frac{\text{full rot}}{s}\right)$, according to the units of the angular trajectories. This velocity can be translated into an angular velocity via

$$\omega_{\text{angular}} = 2\pi\omega_{\text{rot}} \times \frac{180\,^\circ}{\pi} = \omega_{\text{rot}} \times 360\,^\circ\,. \tag{7.3}$$

Typical values range from 0 to $\sim 0.2\,\frac{\text{full rot}}{\text{s}} \widehat{=} 72\,\frac{^\circ}{\text{s}}$. The TT measurements with $p_{\text{reject}} < 99.9\%$ also typically have small velocities of only up to $\sim 0.05\,\frac{\text{full rot}}{\text{s}}$. This threshold separates most of the measurements with high and low rejection probability. If the TT measurement does not show a clear linear trend, the linear fit is not suitable to describe the behavior of $\varphi(t)$. However, in these cases, also the total covered angle is small compared to directed angular trajectories allowing to incorporate both into comparisons.

The velocity is related to the asymmetry between ascending and descending transitions

$$A = \frac{n_{\text{asc}} - n_{\text{desc}}}{n_{\text{asc}} + n_{\text{desc}}} = \frac{n_{\text{asc}} - n_{\text{desc}}}{n_{\text{tot}}}\,, \tag{7.4}$$

following the relation

$$\omega_{\text{rot}} = \frac{1}{3} \times A \times \frac{n_{\text{tot}}}{t_{\text{tot}}}\,, \tag{7.5}$$

assuming that ω_{rot} (extracted from a linear fit) is the same as the slope of a line between $\varphi(0) = 0$ and $\varphi(t_{\text{tot}}) = \frac{2\pi}{3}(n_{\text{asc}} - n_{\text{desc}})$, which is a valid assumption as angular trajectories always start at $\varphi(0) = 0$ and exhibit linear behavior. The factor $\frac{1}{3}$ comes from the switching rate being units of numbers of switching events, while ω_{rot} is in units of $\frac{\text{full rot}}{s}$.

Both velocity and asymmetry are a measure of how large the surplus of the transitions in one direction is over the other. For the asymmetry, this is clear. The velocity incorporates this surplus in a steeper increase of the angle in the same amount of time at the same switching rate. This already illustrates the difference between the two.

Because the asymmetry is normalized to the total amount of transitions, it cannot distinguish between TT measurements with large or small numbers of transitions. The velocity, however, depends also on the amount of transitions that occurred during the complete time of the measurement. Therefore, it can also yield a small velocity, even if the total amount of transitions is small, while the asymmetry could still be large in this case.

To illustrate this, consider two TT measurements. Both have the same asymmetry, but in the same time, one has a small n_{tot} and the other's is large. The first measurement would have a small velocity, as the amount of full rotations per second is small, while it would be large for the latter. Hence, the velocity combines the asymmetry with the total amount of transitions.

Both of these figures are important in the description of the rotation. As it becomes evident in Subsec. 7.4.3, upon looking at both quantities together with the number of transitions, the cause of varying velocities can be disentangled. This helps in interpreting the change of the rotation with i.e. the bias voltage.
Summing up, the velocity adds a figure of merit describing the observed rotation of the functional groups of Tpdx-G. It gives a physical meaning to the observed switches in current levels and provides a means of describing how fast the rotation is. To verify that this type of evaluation is valid, a commonly used method from the description of Brownian motion is applied, namely the mean square displacement.

7.3.3 MSD Plots

As described in Subsec. 2.5.2, the mean square displacement (MSD) is a widely used technique for the evaluation of diffusion processes, e.g. of cells or particles in liquids. Looking at Fig. 2.9, one finds that a plot of the MSD versus the lag time is a suitable means of distinguishing between random and directed motion. Moreover, Eq. 2.20 allows for an extraction of the diffusion constant and the velocity of a driven motion. Because a driven motion superimposed on a random movement of the functional group is expected here, the technique should also be applicable.
Translating the quantities introduced in Subsec. 2.5.2 to the rotation of Tpdx-G functional groups, the particle position r corresponds to the angle φ and the number of dimensions is $n_{\text{dim}} = 1$.
For the determination of the diffusion constant and the velocity, the plot of the MSD as function of the lag time τ is needed. Due to the discrete nature of the measurement, the calculation of the integral in Eq. 2.19 is also done discretely. It is calculated as

$$\text{MSD}(\tau_n) = \langle \varphi^2 \rangle (\tau_n) = \frac{1}{N_{\text{tot}} - n} \sum_{i=0}^{N_{\text{tot}}-n-1} \left[\varphi(t_i + \tau_n) - \varphi(t_i) \right]^2 , \qquad (7.6)$$

where i enumerates the data points of the measurement, t_i is the acquisition time of the ith data point, τ_n is the lag time with $\tau_{n+1} - \tau_n = t_{i+1} - t_i$, i.e. in the same time steps as the measurement, N_{tot} is the total amount of pixels in the measurement and n enumerates the lag times.
To illustrate the calculation of Eq. 7.6, consider one lag time τ_n. All squared distances of φ with a delay of τ_n between them are averaged over the whole range of the measurement. The summation ends, when $t_i + \tau_n = t_{N_{\text{tot}}}$. This implies, since τ_n increases, that the amount of points over which the MSD is averaged decreases. In the extreme case of τ_0, the MSD is the average of the maximum of N_{tot} elements, while for the other extreme of $\tau_{N_{\text{tot}}-1}$ the averaging only incorporates one value.
This directly points to a problem for the fitting process. For larger values of τ_n, the

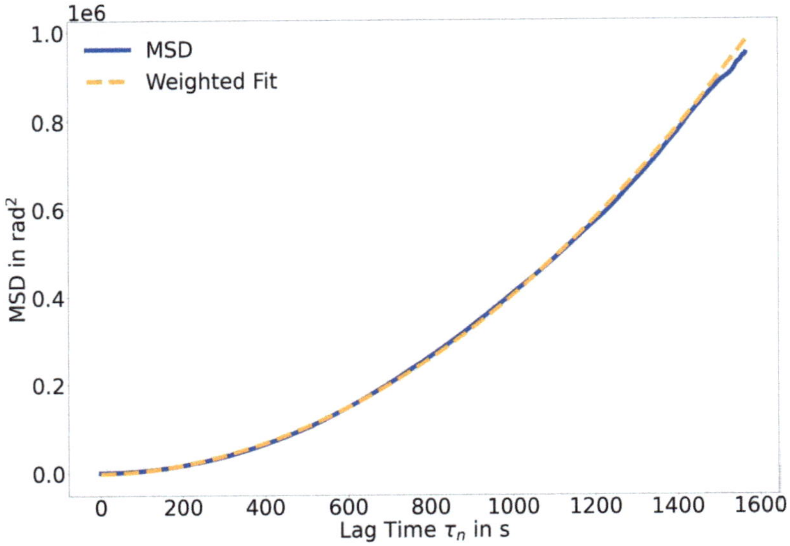

Figure 7.7: MSD plot of TT measurement. The MSD clearly exhibits quadratic dependence on the lag time. The weighted, quadratic fit matches the data very well, suggesting directed motion (compare Fig. 2.9).

averaging process is done with fewer elements and becomes increasingly unreliable. Consequently, it leads to larger variation in the MSD values for large lag times. As a solution, the points in the MSD plots are weighted for the fitting algorithm with the amount of elements in the sum. As a result, more emphasis is put on the curve fitting to small values of τ_n, rather than fitting to large τ_n.
To further simplify the fitting process, the contribution of anomalous diffusion (compare Eq. 2.20) was assumed to be negligible. The problem with the fitting of an anomalous diffusion is the unknown exponential often causing the fit not to converge. Moreover, the offset term MSD_0 was added to the fit function. This offset is related to uncertainties in the measurement [87]. Finally, with the angular velocity ω_{MSD}, the fit function reads

$$\mathrm{MSD}(\tau_n) = (\omega_{\mathrm{MSD}}\tau_n)^2 + 2D\tau_n + \mathrm{MSD}_0\,. \tag{7.7}$$

As is shown in Fig. 7.7, showing a MSD plot for a TT measurement with high asymmetry and the performed square fit, the assumption of negligible anomalous diffusion is justified. The MSD clearly depends quadratically on the lag time, suggesting a dominant contribution of directed motion. Both the linear term and the offset contributions lose significance quickly compared to the squared term with increasing τ_n and do not seem to have a large influence on the quadratic dependence.

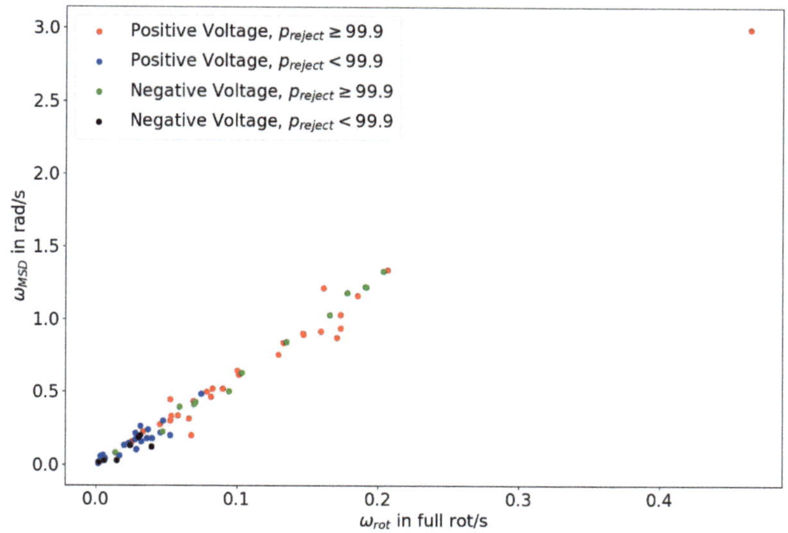

Figure 7.8: Plot of the angular velocities extracted from the MSD as a function of those from angular trajectories. As expected, they depend on each other linearly, indicating that both evaluation methods are viable.

In the exemplary measurement shown in Fig. 7.7, the quadratic fit parameter is $\omega_{\text{MSD}}^2 = 0.39\,\frac{\text{rad}^2}{\text{s}^2}$ which corresponds to $\omega_{\text{MSD}} = 0.62\,\frac{\text{rad}}{\text{s}} \widehat{=} 0.099\,\frac{\text{full rot}}{\text{s}}$. The linear fit parameter is $D = 7.07\,\frac{\text{rad}^2}{\text{s}}$ and the offset is $\text{MSD}_0 = -1903\,\text{rad}^2$. It should be mentioned, that specifically the offset is no usable quantity, as the error from the fit is more than an order of magnitude larger than the actual value. The diffusion constant is a more reliable quantity with reasonable errors. However, in the scope of this work, it will not examined in more detail.

The velocities extracted from fits to the MSD plots range from 0 to $\sim 1.5\,\frac{\text{rad}}{\text{s}} \widehat{=} 0.24\,\frac{\text{full rot}}{\text{s}}$, in acceptable agreement with the range of velocities extracted from the angular trajectories.

For a more detailed comparison, Fig. 7.8 shows a plot of the angular velocities taken from the same measurements using the two methods of calculating the velocity of the rotation. Here, ω_{MSD} is plotted as a function of the corresponding ω_{rot}.

It becomes apparent, that they depend linearly on each other, as it is expected. If they describe the same property of the rotation, the slope of a linear fit should yield the conversion factor between the two quantities. The slope is calculated to be $6.4\,\frac{\text{rad}}{\text{full rot}} \approx 2\pi$, in good agreement with the conversion factor between radians and full rotations.

This linear dependence spans over all types of measurements, including low and high rejection probabilities, as well as positive and negative bias voltages. The good

agreement and linear behavior indicate, that the different evaluation methods describe the same property and can be used equivalently. As the calculation of the angular trajectories is faster than for the MSD, this method is preferred.
The evaluation using the MSD also supports the observation of directed motion. Because the MSD is a commonly used technique, the good agreement gives a more sound foundation to the claim of directed motion.
To conclude, this section provides three different pieces of evidence suggesting directed rotation of the functional group of Tpdx-G. Moreover, the figures of merit describing the rotation and their calculation from TT measurements, namely the rejection probability and the angular velocity, are now clarified. These are the necessary tools to study the rotation on a more fundamental level, investigating its dependences on the various parameters of the measurement.

7.4 Dependences of the Rotation

The description of the important properties of the rotation and their calculation lays the foundation for the investigation of the dependence of the rotation on measurements parameters such as the bias voltage, the current and the tip position. These dependences are discussed in detail in this section.
It begins with plots of the characteristics for the complete set of evaluated measurements across numerous different molecules in Subsec. 7.4.1. However, these do not prove useful for the observation of overall trends in the measurements. That is why the following subsections present measurement series, in which only one parameter is varied and all others are kept constant.
This implies, that the TTs of each series have to be taken one after the other, as the exact same relative position of tip and molecule turns out to be a critical parameter and is almost impossible to find in separate measurements. It poses restrictions on the size of the data set of the dependences, because every data point corresponds to a TT measurement, usually taking 10 min. Even at low temperatures, the drift of the tip position on such small spatial scales is not negligible on these time scales.
Bearing this in mind, it is still possible to measure the dependences, helping understand the cause of the rotation. They present a very strong dependence on the tip position, both laterally and vertically. Most interestingly, the asymmetry of the rotation depends non-monotonically on the bias voltage. It shows a peak-like shape, indicating resonant behavior for a certain voltage. Lastly, the dependence on the bias polarity is interesting with regard to the CISS effect, possibly causing the rotation. Reversing the voltage polarity should lead to a inversion of the rotation direction, if the CISS causes the rotation. So, having measurements at both polarities at the same lateral tip position might give insight into the effect driving the motion.

7.4.1 Scatter Plots

The description and calculation of the figures of merit of the rotation is the foundation on which the investigation of the underlying effect of the directed motion is built. Velocity, asymmetry, rejection probability and switching rate ($\frac{n_{\text{tot}}}{t_{\text{tot}}}$) are the main characteristic values.
To get insight into the cause of the directionality, the dependence of these with respect to other quantities, such as bias voltage and average current or between each other have to be studied. The most intuitive way is to simply plot these values for all single TT measurements evaluated (89 in total). These TTs are a subset of all data acquired, being filtered manually by choosing TTs that show switching between three metastable current levels on time scales suitable for STM. They can give an overview of overall trends in the complete data set. These plots will be called scatter plots in the following.
Fig. 7.9 shows four examples of such scatter plots. The color coding separates the data into four categories, distinguishing high and low rejection probability in the binomial test (threshold $99.9\,\%$), as well as positive and negative bias voltage. This makes it easier to see whether a certain type of TT measurement exhibits a specific overall trend.
All scatter plots shown in Fig. 7.9 exhibit a very large spread in the data. For all of them, it seems that the spread of the plotted figure of merit exceeds the range of voltages and currents for which it is possible to perform and evaluate a TT measurement, meaning that overall trends cannot be captured because of the limited available parameter space for TT measurements. Bearing in mind that the plots cover various different molecules, relative tip-molecule positions, tip shapes etc., it seems plausible that the spread is large. This already indicates that the tip position relative to the molecule plays a major role in how the molecule rotates. A more detailed investigation of the dependence of the rotation specifically concerning the lateral tip position can be found in Subsec. 7.4.2.
Fig. 7.9a shows a scatter plot of the average current as a function of the bias voltage. The current is calculated as the average of the center of the current distribution of levels 2 and 0. It becomes apparent, that the bias voltages for which TT measurements show switching between three metastable current levels on time scales suitable for STM is limited to $\sim 1.6 - 2.0\,\text{V}$ and the corresponding currents go as high as $\sim 400\,\text{pA}$. However within this range, no clear dependence of all TTs can be observed. They are distributed in this window, without following a defined trend. Still, the observed behavior that TT measurements at negative bias voltages exhibit higher current values can be visualized in this scatter plot. While most measurements at positive voltage are distributed up to currents of $200\,\text{pA}$, the ones at negative voltage range from $200 - 400\,\text{pA}$ to achieve similar switching rates.

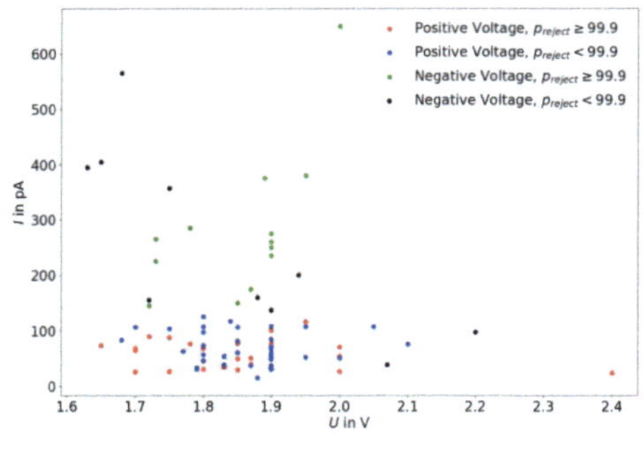

(a) Average current vs bias voltage.

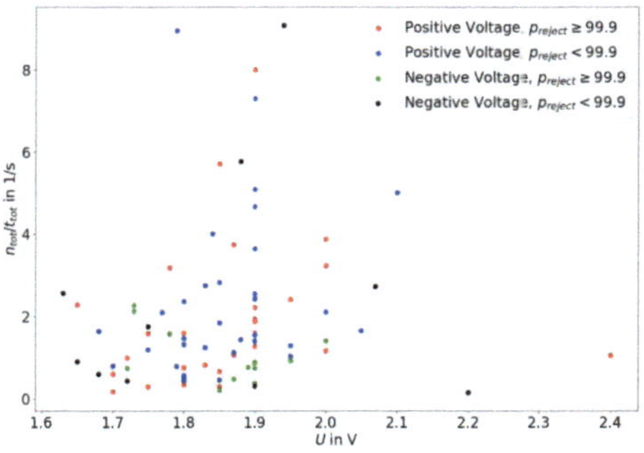

(b) Switching rate vs bias voltage.

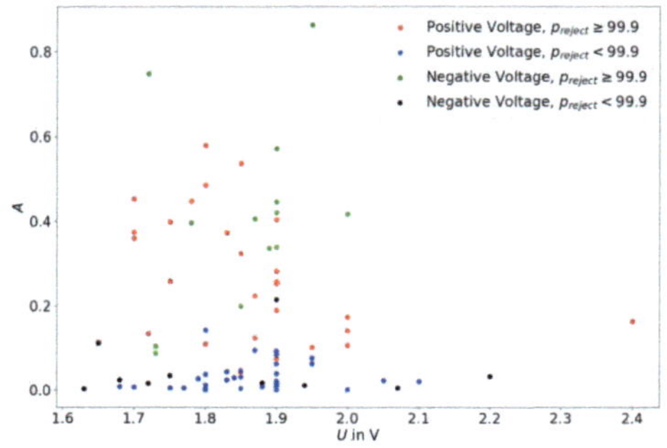

(c) Asymmetry vs bias voltage.

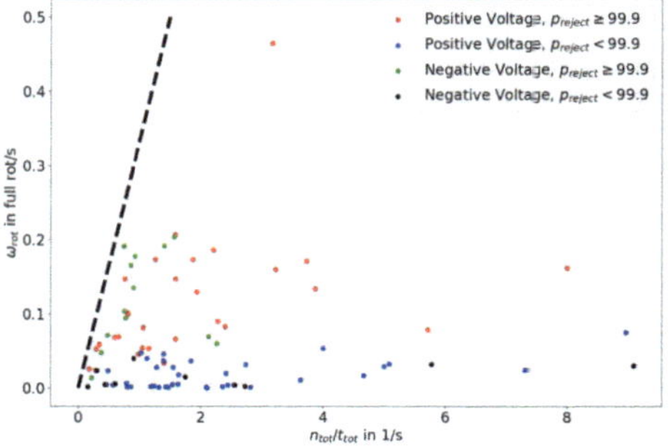

(d) Velocity vs switching rate.

Figure 7.9: Scatter plots of different figures of merit for 89 evaluated measurements. They are separated into high and low rejection probabilities, as well as positive and negative bias voltage. The black dashed line indicates the maximum possible velocity for $A = 1$, according to Eq. 7.5.

It is unclear what causes this observation. An explanation could be, that the orbitals contributing to the transport through the molecule at positive bias are more weakly connected within the molecule and do not lead to as high currents at the same voltage at almost identical tip position. Possibly there are also fewer orbitals contributing to the transport through the molecule at positive bias.

In Fig. 7.9b the switching rate is plotted as a function of bias voltage. This plot is the only one, where arguably an overall trend can be observed. The switching rate increases with the applied bias voltage, holding for all types of measurements. However, even here, the spread is relatively wide and it is not sensible to try and extract numerical values from this correlation. Moreover, this plot reveals that a large switching rate does not imply a large rejection probability. In fact, the measurements with large p_{reject} are distributed over the whole range of switching rates. This is an indication that the random switching and the directed drive are based on two different effects.

A similar observation can be made in the scatter plot of the asymmetry as function of bias in Fig. 7.9c. Here, TT measurements of both positive and negative voltage with high p_{reject} are found in the whole range of bias voltages. So in this scope, also the bias voltage does not seem to be one of the governing parameters to induce directed rotation. This changes, however, looking at measurements at a fixed tip position, presented in Subsec. 7.4.3.

What can be learned from this scatter plot is that a large asymmetry, in combination with a large enough number of events, also implies a large rejection probability. This behavior is expected, but it validates that the figures of merit extracted are sensible and consistent with each other. Empirically taking the values from Fig. 7.9c, the rejection probability threshold of $99.9\,\%$ sets a threshold in the asymmetry of ~ 0.1, with typically $n_{\text{tot}} > 100$. As an estimation a binomial test can be performed with the average $n_{\text{tot}}^{\text{avg}} = 1483$ of the measurements included. Assuming an asymmetry of 0.1, the rejection probability is $99.9926\,\%$. This value is comparable to the actual threshold rejection probability, bearing in mind the rough estimation of the threshold asymmetry.

The last scatter plot shown here is of the velocity plotted as function of the switching rate. Here, the maximum possible velocity is plotted according to Eq. 7.5 as a black dashed line. Also in the velocity, a separation between low and high rejection probability can be observed for low and high velocities, as expected. For the higher velocities, TTs at positive and negative bias voltage show different behavior. Much like in the other scatter plots, the velocity spreads over the whole observable range. There is no clear trend visible. For negative voltages, the velocity seems to increase with increasing rate. However, there are also measurements at higher switching rates with low velocities. Also, there are fewer measurements at negative bias, so the lack of measurements with lower velocities might be due to missing data. The spread in all other values makes this scenario most probable, especially since the equivalent at positive bias also has a large spread.

A way to circumvent the problem of large spread in data is to limit the plotted data to a certain subset. An intuitive way is to plot it for a single bias voltage. This was done in Fig. 7.10. Here, the velocity as function of switching rate is plotted only for $U = 1.9\,\text{V}$, because it is the voltage with the most TT measurements. Even for this single bias voltage, the velocities for high rejection probabilities range from $\sim 0.03 - 0.18\,\frac{\text{full rot}}{\text{s}}$ and the switching rates from $\sim 0 - 2\,\frac{1}{\text{s}}$.

Comparing this to the overall plot in Fig. 7.9d one can see that these are almost the full ranges of velocities and switching rates for the evaluated measurements. This means that even for a single bias voltage, the spread in data is almost the whole range of observed values.

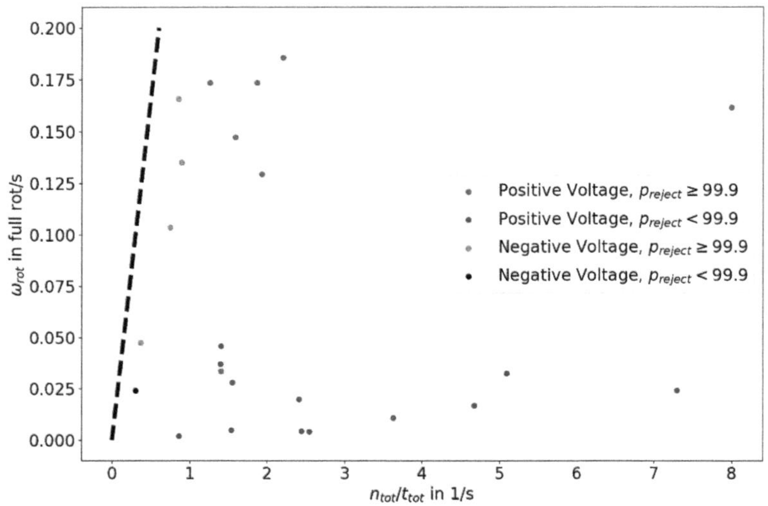

Figure 7.10: Plot of velocity vs switching rate for measurements at $U = 1.9$ V. Even for one bias voltage, almost the whole range of observed values in covered. The black dashed line again indicates the maximum possible velocity for $A = 1$.

All these observations point towards the fact that a parameter, which is not incorporated in the systematic description so far plays a major role in the directed rotation. This is the relative tip-molecule position. This is supported by TT measurements performed on a grid on a single molecule at a fixed bias voltage (Subsec. 7.4.2), as well as current-dependent measurements (Subsec. 7.4.3).
In conclusion, even though the scatter plots do not yield the visualization of overall trends in the complete data set, as the spread is too large, they are still useful. With their help, it is possible to better quantify observed behavior such as the increase in switching rate with bias voltage. Also they are in agreement with the behavior expected from the description of the figures of merit, validating the methods of evaluation. Most importantly, they show that taking measurements from various molecules and relative tip-molecule positions creates a too large variation in the results for comparison. The cause of this variation becomes more clear by investigating the dependence of the rotation on the lateral tip position.

7.4.2 Grid Measurements

It becomes evident from the scatter plots in Subsec. 7.4.1, that the variation in the characteristics of the rotation is very large when comparing different molecules, tip position etc. Here, a more detailed investigation of the dependence on the lateral tip position above the molecule is presented.

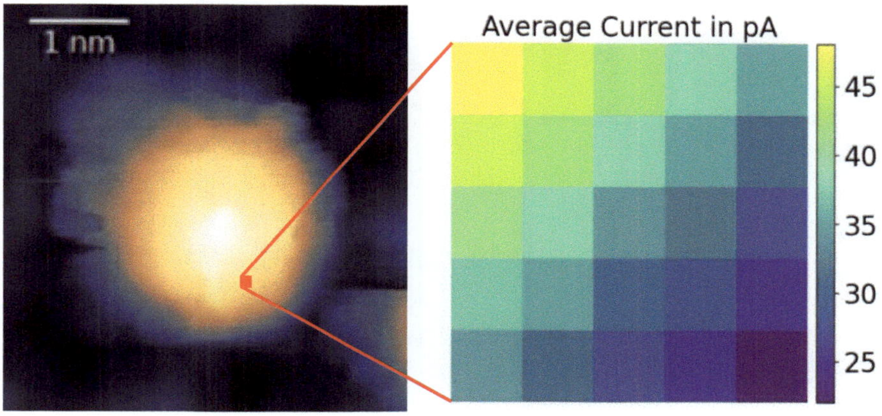

Figure 7.11: Constant current image of the single molecule on which the grid measurements are performed. Superimposed in red is the area of the 100 pm grid. The right side shows the average current of the measured TTs at every pixel. The smooth transition indicates no major, abrupt changes during the grid measurement.

This dependence is measured by fixing the the bias voltage (1.93 V) and the z-position, and slightly changing the lateral tip position. The most straightforward way is to measure on a grid of tip positions. For this purpose, a tip position, where the rotation works nicely is chosen and a series of 5 minute TT measurements is taken on a 5×5 pixel grid surrounding this position. The TT measurements are evaluated individually and their results are mapped according to the lateral tip position.
Two grids are measured on the same, single molecule and similar lateral tip positions. Neither the bias voltage, nor the z-position is changed between the two measurements for reasons of comparability. One is performed on a grid with an edge length of 100 pm and the other with 50 pm, comprising the bottom left quarter of the larger grid. For a comparison of the size of the grid with the size of the molecule, Fig. 7.11 shows the square of the 100 pm grid measurement superimposed on the constant current image.
The same figure also depicts the evolution of the average current depending on the lateral tip position. The average current is calculated by taking the average of the current values of states 0 and 2. The evolution is smooth, meaning that there was no tip change and no major, abrupt change in the configuration of the molecule during the measurement. Moreover, it decreases from the top left corner to the bottom right corner, as expected from the topography. This confirms that the TT measurements of the grid are comparable and consistent.
A first impression of the spatial evolution of the rotation is given by the rejection probability. The corresponding heat map can be found in Fig. 7.12a. It shows that, on a scale of 100 pm, the probability of the rotation being directed goes from being low

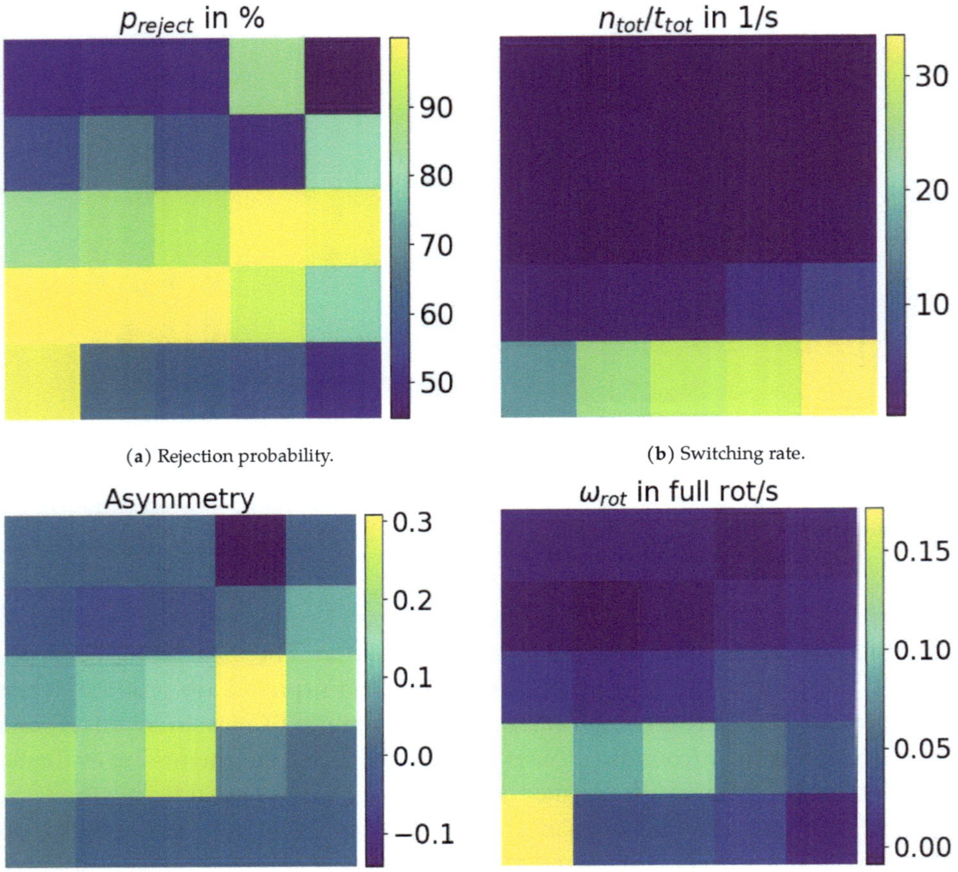

(a) Rejection probability.

(b) Switching rate.

(c) Asymmetry.

(d) Velocity.

Figure 7.12: Heat maps of important characteristics of the rotation depending on lateral tip position in a grid with 100 pm edge length. Bias voltage and z position of the tip are held constant during the measurement.

at the top left, via a almost certain directionality, back to being low at the lower right. The area of high rejection probability forms a stripe, roughly at constant radius from the center of the molecule in Fig. 7.11. Although the rejection probability does not provide insight into the causes of the changing directionality, it indicates the high sensitivity of the rotational switching to the lateral tip position.

In fact, looking at the TT measurements itself already gives an idea of why the rejection probability changes so drastically. Figure 7.13 shows three cutouts of different duration of TT measurements taken at the top left, center and bottom right of the grid. It is clearly visible, that Fig. 7.13a only features two states and can therefore not show any directionality. The measurement in Fig. 7.13b does show three states, they are all occupied frequently and the switching rate of all six possible transitions is reasonable. In contrast, the measurement at the bottom right of the grid (Fig. 7.13c) shows a very high overall switching rate. However, the transitions occur almost only between two of the three possible states, effectively reducing the asymmetry and the surplus of rotations in one direction by increasing the overall number of transitions. Transferring this behavior to the interpretation presented in Subsec. 6.2.3, it means that the head group is moving back and forth between two of the three states, but rarely completes a full rotation.

To show that this behavior is not restricted to single TTs but is an overall trend in the grid, it is useful to take a closer look at the heat maps of the switching rate and asymmetry. The heat map of the switching rate (Fig. 7.12b) shows a strong increase in the switching rate going from the top left to the bottom right. This corresponds well to the TT measurements shown in Fig. 7.13 and shows that the transition from small to large switching rates is smooth.

Taking the evolution of the average current into account, the increase in switching rate occurs for a decrease in current. At first sight, this is unexpected, as higher currents in general should lead to higher probability of interaction with the molecule, therefore to higher probability of excitation and thus to higher switching rates. However, in a more mechanical point of view, the behavior can be rationalized. For higher currents, the tip is closer to the head group, increasing hindrance effects, effectively reducing the number of switching events. In other words, the tip 'gets in the way of the rotation' with increasing currents. For smaller currents, the head group is more free to rotate and the switching rate increases.

For the asymmetry, the evolution looks very different. It has a maximum approximately in the center of the grid and decreases in all directions, most significantly to the top left and bottom right. As seen in Fig. 7.13a, the top left measurements do not feature a third state, not allowing for any asymmetry. In the other extreme case at the bottom right, the overwhelming amount of transitions between states 0 and 1 dominates the number of total transitions. These transitions do not create asymmetry, as there is just switching back and forth, creating equivalent numbers of transitions increasing and decreasing the current. The difference between these ascending and descending transitions becomes small compared to the total amount of transitions. Only in the center of the grid, there is a reasonable asymmetry. All three states are assumed frequently and all transitions take place in comparable amounts, leading to a reasonable asymmetry.

Finally, to achieve higher velocities (Fig. 7.12d), both asymmetry and switching rate have to be high enough, which implies that the position of maximum asymmetry or switching rate is not necessarily the position of maximum velocity. Comparing the heat map of the velocity to the switching rate and asymmetry, the position of maximum velocity is at neither of their maxima. It is at a position where both are

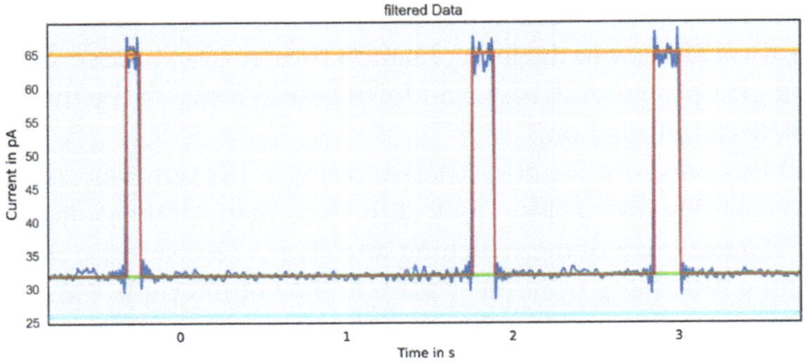

(a) TT at top left. Only two states are observed excluding any kind of directionality.

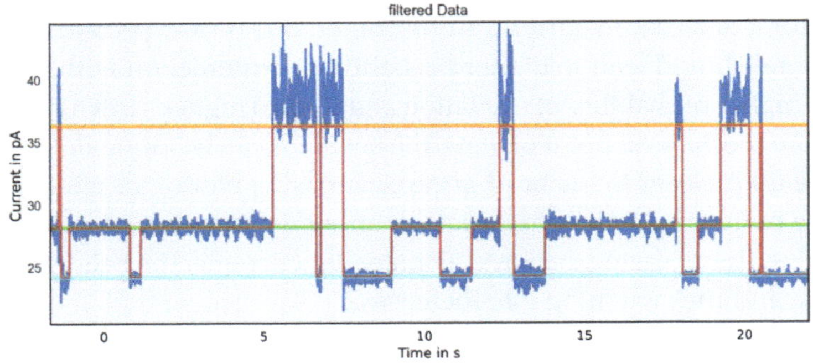

(b) TT at center. Switching rates and proportion of the different transitions is reasonable.

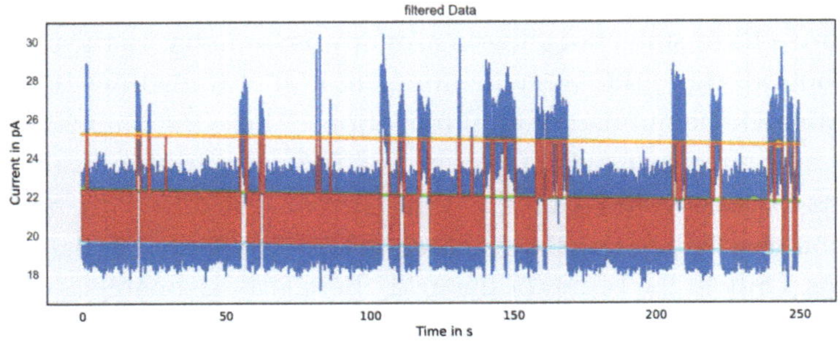

(c) TT at bottom right. One transition is clearly favored, overshadowing directed rotation.

Figure 7.13: Cutouts of different duration of three TT measurements in the 100 pm grid.

reasonable. Towards the top, the switching rate becomes too low and towards high switching rates, the asymmetry decreases rapidly. This leaves only a small area on the grid with sensible velocities.
For better lateral resolution, the same type of measurement is performed on a 50 pm grid in the bottom left quarter of the previous grid. The results are shown in Fig. 7.14. Again, the evolution of the average current (see Fig. 7.14a) is smooth, indicating no tip changes and ensuring comparability of the TT measurements.
Here, the evolution of the current is smooth and all other characteristics are in very good agreement with the larger-scale grid. If compared with the 100 pm grid, the evolution of the figures of merit presented with the lateral tip position, as well as their numerical values match very well. On this scale, however, all TTs show three states and reasonable velocities. This means that here, also the switching rate and asymmetry are reasonable over the whole grid.
The increased lateral resolution reveals, that the changes observed on the larger scale are also smooth at higher resolution, underlining that the observed trends actually reflect the rotational behavior of the head group and are not coincidental or random. Moreover, even on this scale, the velocity is not homogeneous, meaning that there is still a variation in the interaction between the tip and the sample which significantly changes the velocity of the rotation. Within a lateral distance of 25 pm, the velocity is roughly halved. This explains, why the scatter plots exhibit such large variations. When observing different molecules, it is impossible to find the exact same configuration of tip and molecule with a precision of 25 pm. Because the variations, even on very small scales, are so large, the variations when studying different molecules are also significant. Therefore, only measurements taken at the exact same molecule and fixed lateral tip position can be compared directly in order to study the influence of the applied bias voltage and tunneling current.
The persistence of the significant dependence of the velocity on the lateral tip position, even at length scales of a few 10 pm, points towards a very short-ranged and strong interaction between the tip and the molecular head group, which is fulfilled i.e. by chemical forces. Electrostatic interaction, for example, would not exhibit such a strong distance dependence. It can therefore be compared to mechanical interactions on the macroscopic scale. This would also explain the observation of an increasing rate with increasing tip–molecule distance and the observation of only two states at higher currents, as mentioned before, allowing for an intuitive understanding of the behavior seen in Fig. 7.13.
The populations of the states 0 to 2 can also be plotted using the data from the grid measurements, which is shown in Fig. 7.15 for the larger grid. The population of a state is the time, for which the current assumes the corresponding state, summed for the complete measurement. Firstly, the heat maps show, that state 0 is not assumed in the three pixels at the top left. Secondly, the areas in which states 0,1 and 2 are

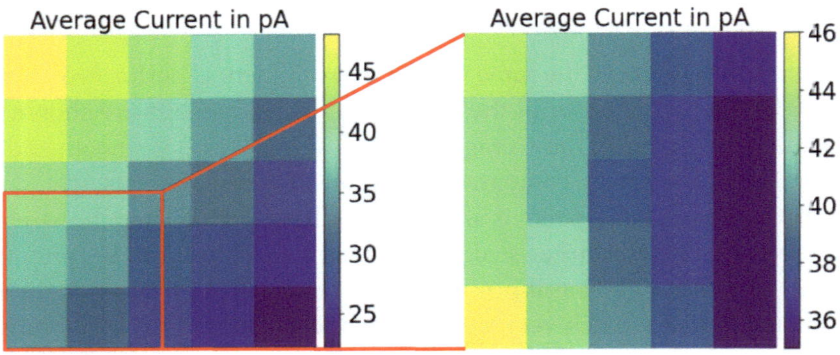

(a) Average current for the 50 pm grid (right) and its location within the 100 pm grid (left).

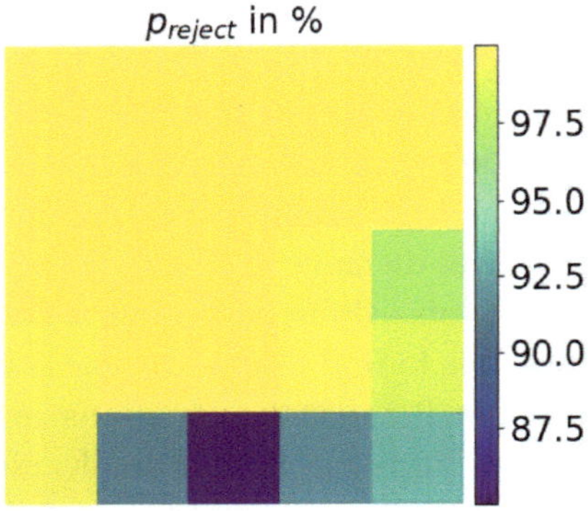

(b) Rejection probability.

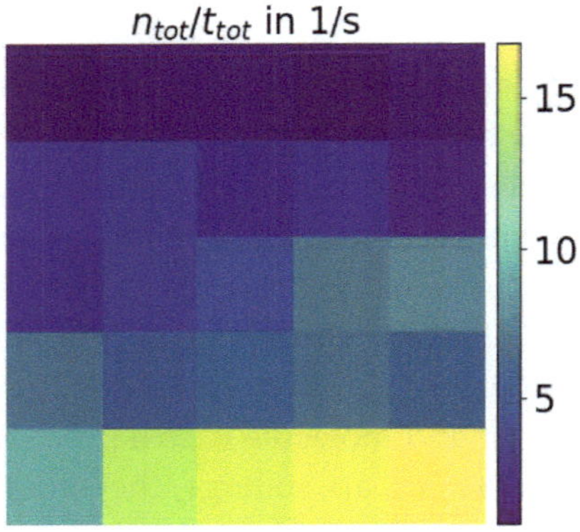

(c) Switching rate.

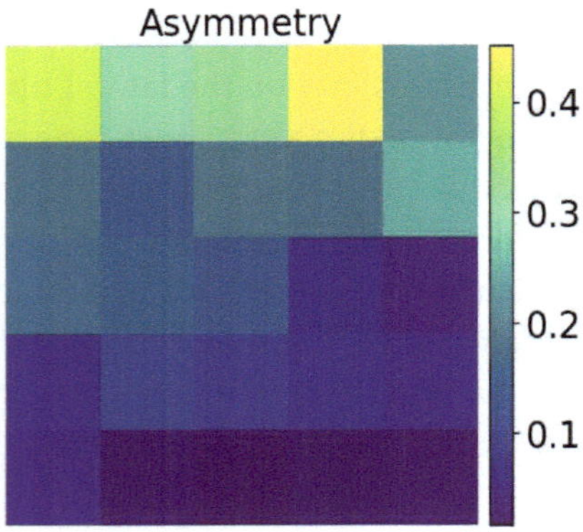

(d) Asymmetry.

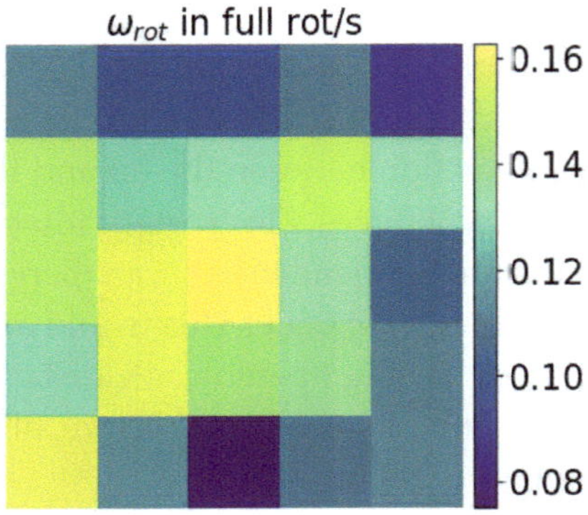

(e) Velocity.

Figure 7.14: Heat maps, acquired as for the 100 pm grid measurement, for a grid with 50 pm edge length in the bottom left quarter of Fig. 7.12.

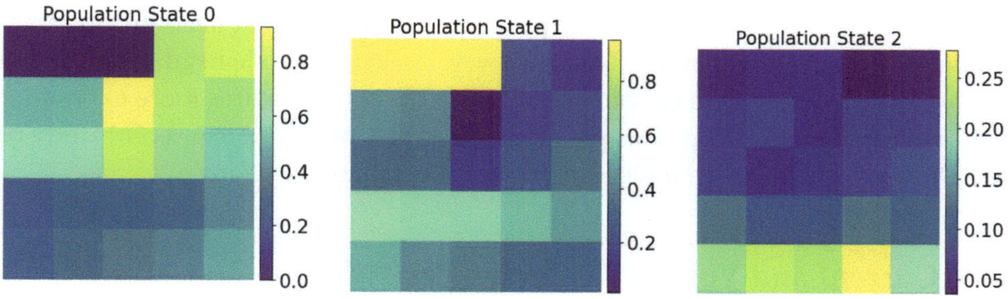

(a) State 0 population. (b) State 1 population. (c) State 2 population.

Figure 7.15: Heat maps of the population of states for the larger grid measurement. They show a spatial variation in the favored state.

assumed the longest vary spatially. Specifically, state 0 is mostly assumed at the top left, state 1 towards the bottom (excluding the three pixels at the top left, as they are distorted, because state 0 is not assumed) and state 2 at the bottom line. This explains that the single states cannot be imaged, but only the superposition of all states are observed, as they go from being assumed frequently to almost not being assumed on a length scale of $< 100\,\mathrm{pm}$. Also, the spatial distribution of favored states is an indication, that the observed movement is a spatial movement of the head group and not a reconfiguration of another part of the molecule.
Including the absolute changes in current between the three levels observed in the TT measurements, the spatial variation of the apex of the molecular head group corresponding to the directed rotation can be estimated. Here, the current varies between $\sim 10\,\mathrm{pA}$ and $40\,\mathrm{pA}$ (see i.e. Fig. 7.13b). Applying the rule of thumb mentioned in Subsec. 2.1.1, that a change of one order of magnitude in current corresponds to a change of $1\,\text{Å}$ in tip–sample distance, the observed variation in tip–molecule distance is on the order of a few $10\,\mathrm{pm}$, but less than $100\,\mathrm{pm}$. This would not agree with switching of the inclination of the complete head group between the three spaces between or on the foot structures (as previously sketched in Fig. 7.2, described in Subsec. 6.2.2 and visible from the simulations shown in Subsec. 7.5.1). It would make a rotation of the head group around the axis defined by its backbone more probable. A more detailed analysis is presented in Sec. 7.5.
In conclusion, the grid measurements have shown that the dependence of the rotation on the lateral tip position is significant, even down to a scale of a few $10\,\mathrm{pm}$, and a very specific tip position is needed to achieve sensible velocities. It explains the large spread observed in the scatter plots presented in Subsec. 7.4.1. This is due to a balance between the distribution of high asymmetry and switching rate over the measurable area. These large variations on small scales only allow for comparing measurements taken at the same lateral tip position.

The crucial role of the STM tip on the properties of the rotating head group can be qualitatively understood in the following way: As the rotatable head group points away and is separated, to a certain extent, from the substrate, it is more exposed to the influence of the STM tip. This phenomenon is also observed in a recent study on the light emission of single molecules in the STM junction [88].

7.4.3 Current and Voltage Dependences

As concluded from Subsec. 7.4.2, the strong dependence on the lateral tip position necessitates keeping it constant to allow for comparison between measurements. Thus, the current and voltage dependences discussed in the following are studied at fixed lateral tip positions.

For the current dependence, the tip position has to be varied perpendicular to the surface due to the nature of the tunneling current. Intrinsically, STM does not allow for a disentanglement of distance and current, so the dependence of the rotation on these parameters cannot be studied separately. By approaching the surface, the tip–molecule distance is reduced, increasing the current and *vice versa*.

To perform appropriate measurements in this context, statistical analysis is needed for every measured z position of the tip. Therefore, a $\sim$ 10 min TT measurement is performed for every z position and evaluated individually. The results are then plotted as function of the average current (of the current levels of states 0 and 2) of the TT measurement. The results of a measurement including 6 different z positions are shown in Fig. 7.16.

The errors are calculated as $\frac{\sqrt{n_{\mathrm{tot}}}}{t_{\mathrm{tot}}}$ for the switching rate and $\frac{2}{n_{\mathrm{tot}}}\sqrt{\frac{n_{\mathrm{asc}}n_{\mathrm{desc}}}{n_{\mathrm{tot}}}}$ for the asymmetry, according to the propagation of errors (variance formula) and assuming the errors of numbers (i.e. n_{tot}) are their respective square roots ($\sqrt{n_{\mathrm{tot}}}$).

The switching rate (Fig. 7.16a) exhibits a similar behavior as for the lateral tip position dependence. The switching rate at the lowest current ($25\,\frac{1}{\mathrm{s}}$) is comparable to the maximum of the grid measurement ($30\,\frac{1}{\mathrm{s}}$). For increasing currents, the switching rate decreases very rapidly, reaching almost zero by approaching only 10 pm, corresponding to a change in average current of 5 pA. In case of the grid measurements, the switching rate goes from the maximum value to almost zero at a distance of 50 pm, which is in a similar range as for the lateral tip position, keeping in mind that the measurements are performed on different molecules.

The comparability of the dependence on the lateral and z position of the tip emphasizes the validity of the argument presented in Subsec. 7.4.2, because if the dependence originates from a strong interaction, such as chemical forces, it should exhibit a similar distance dependence in all three dimensions.

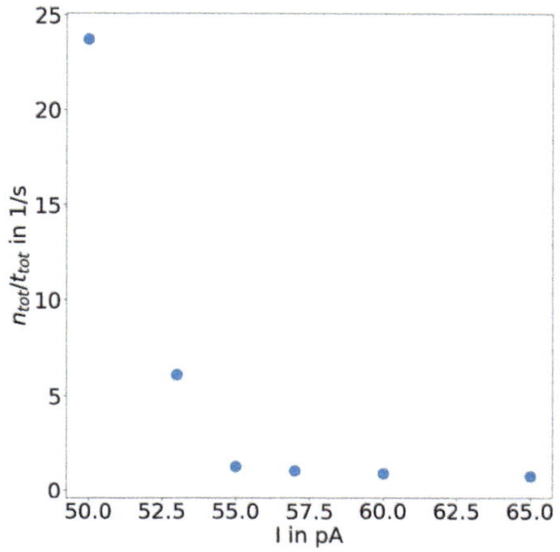

(a) Switching rate. As for the grid measurements, the rate decreases rapidly for increasing currents.

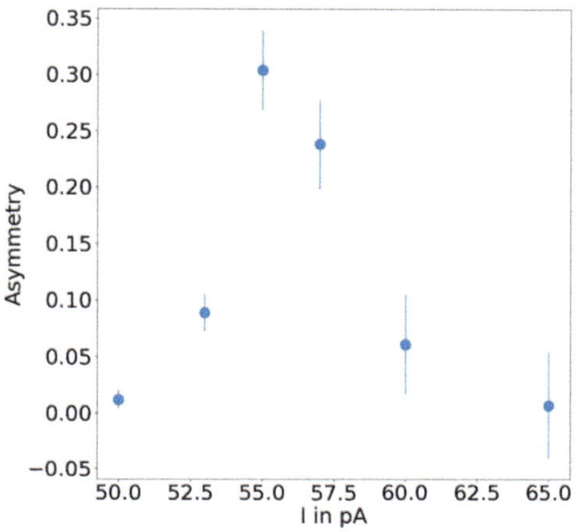

(b) Asymmetry. Similar to the lateral tip position, also small variation in the z position have dramatic effects on the asymmetry.

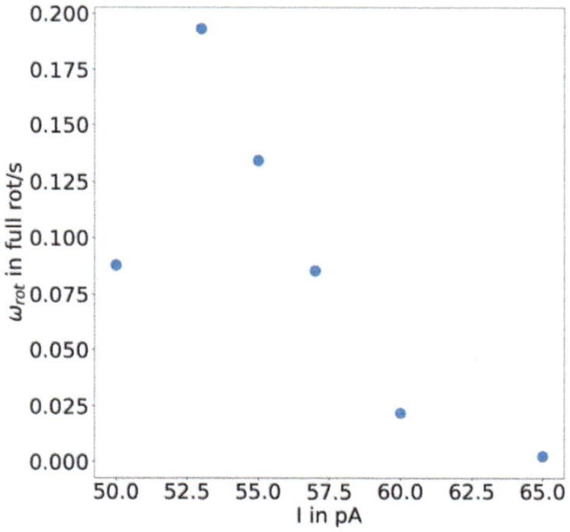

(c) Velocity. To achieve maximum velocity, both reasonable switching rate and asymmetry are needed, as Eq. 7.5 describes.

Figure 7.16: Current dependence of the rotation. For each point, a $\sim 10\,\mathrm{min}$ TT measurement is evaluated. The z position of the tip is changed by $5\,\mathrm{pm}$ between individual points. As for the lateral tip position, only a small window of z positions allows for reasonable velocities.

Similarly, the asymmetry displays a comparable behavior for changes in the z position as for the lateral position of the tip. As Fig. 7.16b shows, the asymmetry features a maximum in the studied z range on a length scale of $25\,\mathrm{pm}$. For the grid measurements this happens on a length scale of $\sim 75\,\mathrm{pm}$. In analogy to the switching rate, the asymmetry exhibits the same qualitative behavior for lateral and z position of the tip. In both cases, the length scales are slightly larger in the lateral tip position dependence, which can be understood by recognizing that the change in tip–molecule distance $\Delta d = \sqrt{\Delta x^2 + \Delta y^2 + \Delta z^2}$ does not scale in the same way for tip displacement in x,y and z, as the head group is inclined at a certain angle with respect to the lot. In the present experiment, the contribution from Δz is expected to be the largest.

The velocity, as is also the case for the grid measurements, needs both the switching rate and the asymmetry to be simultaneously as high as possible to achieve high values. It can be thought of as the product (see Eq. 7.5) of switching rate and asymmetry. If one of the two parameters gets too low, the velocity is also small.

Combined with the grid measurements (Subsec. 7.4.2), the current (or z position) dependence reveals that the directed rotation only works at a very small set of tip positions. It has to be chosen carefully, in all three dimensions, to allow for observation of the directed rotation with reasonable velocity. The small length scales, on which the rotation is detectable, and the similar scales in all three dimensions point towards a short-ranged and strong interaction that defines a potential landscape which allows for the observation of directed rotation.

However, the dependence on the tip position in all three dimensions only defines the corresponding parameter space, in which the rotation is observable. To better understand what drives the rotation, it is crucial to investigate the dependence on the energy of the injected electrons, which is defined by the bias voltage. For that purpose, similar series of measurements are performed as for the current dependence. In these, the tip position is fixed in all three dimensions. TT measurements are taken for varying bias voltages and are evaluated as described in Sec. 7.2. The results of a measurement series are shown in Fig. 7.17.

As expected, the current increases monotonously with increasing bias voltage, with the difference, that the dependence is rather linear than exponential (compare Eq. 2.4). It could be explained by being in a voltage range where the exponential is still comparably flat or on a scale on which the exponential is reasonably well approximated by a linear behavior.

Similarly, the switching rate follows a monotonic, rather linear trend over the covered voltage range. In contrast to the behavior described in Subsec. 7.4.2 and observed in the current dependence, the switching rate now increases with increasing current. It is sensible that the behavior is different, as the tip position is fixed in all three dimensions, excluding a change due to a direct interaction, such as chemical forces.

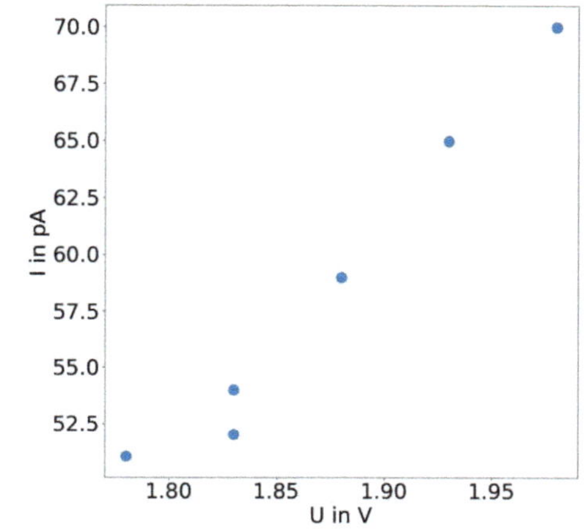

(**a**) Current. As expected, it increases with bias voltage.

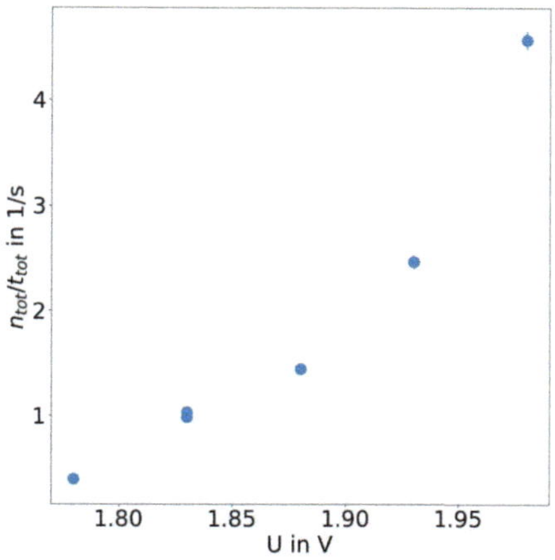

(**b**) Switching rate. It increases with bias and here, higher currents lead to higher switching rates.

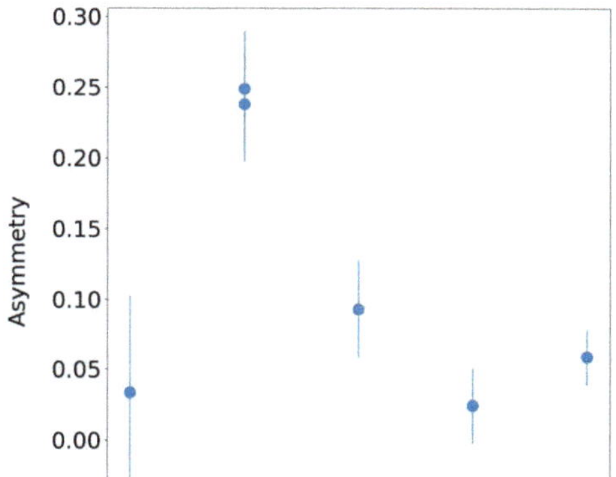

(**c**) Asymmetry. It exhibits non-monotonic behavior with a maximum at 1.83 V.

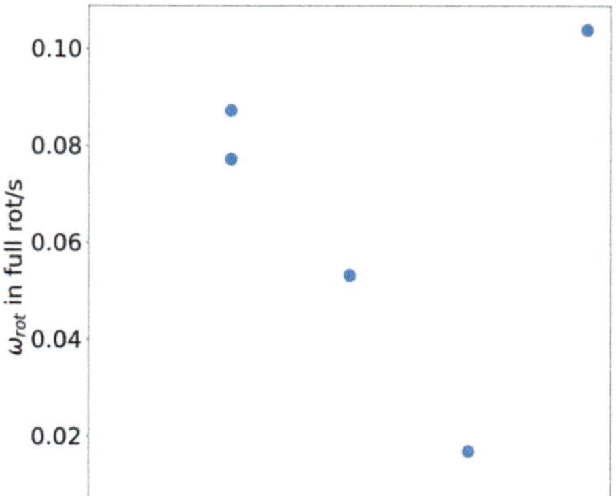

(**d**) Velocity. Again, to obtain higher velocities, both reasonable switching rate and asymmetry are necessary.

Figure 7.17: Bias voltage dependence of the rotation. The tip position is fixed in all three dimensions and the bias voltage is changed. Most interestingly, the asymmetry shows a maximum at a specific voltage.

The tip cannot 'get in the way of rotation' during this measurement, because only the bias voltage is changed and the electrostatic interaction with the head group is rather weak, as the apex of the molecular structure is not expected to have a strong dipolar moment.
If the switching would be caused by inelastically tunneling electrons, a higher current would lead to higher transition rates between states, here specifically resulting in a higher switching rate. In this case, the scaling with the current would have to be according to a power law with roughly $\frac{n_{\text{tot}}}{t_{\text{tot}}} \propto I^5$, implying that 5 electrons are needed for the process, which is unreasonable.
It is more probable that the current is not the crucial parameter in this case and that the switching rate is predominantly excited thermally (compatible with $\sim 10\,\text{meV}$ barrier heights from simulations, Subsec. 7.5.2) and the rate is defined mainly by the bias voltage, which could influence the barrier heights between the metastable states, in turn changing the switching rate.
The most interesting characteristic in this context is the asymmetry (see Fig. 7.17c). It is mostly flat and low throughout the complete observable voltage range, except for one bias voltage. At this specific bias voltage, the asymmetry is more than twice as high as for any other voltage. Such a high local maximum on a rather small voltage scale points towards the crucial role of tunneling into a specific molecular orbital. This would imply that at approximately $1.83\,\text{eV}$ (for all bias dependences at $\sim 1.9\,\text{eV}$), the electrons tunnel into an orbital that has a certain characteristic, which is responsible for driving the directed rotation. To get insight into what this characteristic might be, theoretical calculations of the molecular orbitals are performed an can be found in Sec. 7.5. By comparing the shape of the molecular orbitals and their energies to the observed maximum in the asymmetry, it could be possible to get hints towards the effect causing the directed rotation.
Lastly, the velocity assumes high values at two bias voltages, one at maximum asymmetry and low switching rate and one at low asymmetry and maximum switching rate. This emphasizes once more that the velocity can be thought of as a product of switching rate and asymmetry (compare Eq. 7.5), in that both have to be present to achieve high velocities.
In addition to the bias voltage dependence at positive biases, one measurement series could be performed at negative biases. However, these results have to be considered with some precaution, because four states are observable. It is evaluated, because the transitions between two of the states occur at a higher rate than the other transitions. Counting the two states with fast transitions as one, the measurement can be evaluated and exhibits a very similar behavior to the corresponding measurement at positive bias, qualitatively. The absolute current (Fig. 7.18a) and the switching rate (Fig. 7.18b) increase with increasing absolute bias. Interestingly, also at negative bias voltages, there is an extremal asymmetry at $-1.95\,\text{V}$ (Fig. 7.18c, negative asymmetry

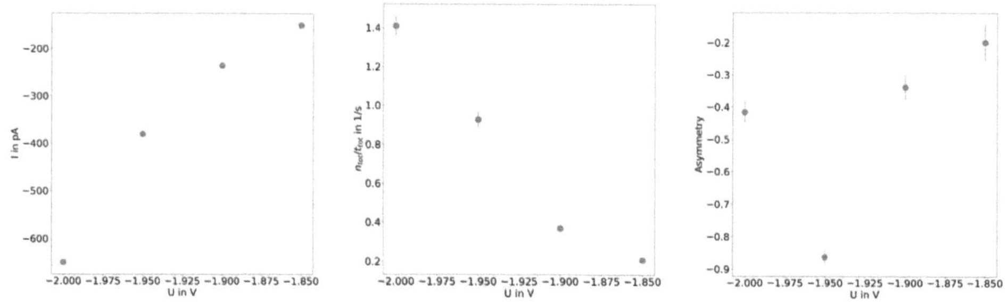

(**a**) Current. (**b**) Switching rate. (**c**) Asymmetry.

Figure 7.18: Bias voltage dependence for negative bias voltages. The behavior is very similar to positive bias voltages, qualitatively. The asymmetry also features an extremum.

because the descending transitions are predominant), suggesting that there is a corresponding molecular orbital below the Fermi energy which also causes directed rotation.

In conclusion, the z position of the tip, as is also observed for the lateral tip position, has a strong influence on whether the rotation can be observed. On very small length scales, there is either a suppressed switching rate, or a suppressed asymmetry resulting in low velocities. This shows that a very specific tip position is necessary to allow for the observation of directed rotation. The bias dependence shows a maximum in the asymmetry, which cannot be explained by the tip position, as it is kept constant throughout that measurement.

Another piece in the study of the cause of directed rotation, described in detail in Subsec. 7.4.4, is to investigate the behavior for different bias polarities.

7.4.4 Voltage Polarity Dependence

In addition to the dependence on the value of the bias voltage value, it is useful to study the directed rotation at the opposite bias polarity. Inverting the bias polarity injects the electrons from the opposite electrode. Here, the bias voltage is applied to the sample, injecting electrons from the sample side at negative bias and from the tip at positive bias. Some effects that could cause directed rotation should invert the preferred direction of rotation if the electrons pass through the molecule in the opposite direction, e.g. the chiral-induced spin selectivity (CISS) effect.

In Subsec. 7.4.3 one measurement series performed at negative bias is presented. In fact, directed rotation is also observed frequently at negative bias voltages, as it can be seen from the scatter plots in Subsec. 7.4.1. It is also already mentioned there, that most TT measurements at negative bias are at much higher currents. The voltages, at which they work, however, are in a similar range as for positive bias. Overall, finding the correct parameters for measuring TTs, with three metastable rotational states and reasonable switching rate, at negative bias turned out to be more challenging than at positive bias. In the measurements evaluated for the scatter plots, 67 are at positive bias and only 23 are at negative bias.

Despite the lower chance of a TT with high rejection probability, it is even possible to find lateral tip positions, at which subsequent TTs at positive and negative bias can be acquired. However, the z position and the value of the bias voltage are potentially different for the two bias polarities. This is somewhat surprising, considering the strong dependence of the observability of directed rotation on the tip position presented in Subsec. 7.4.2 and 7.4.3. It shows that there is an effect, depending on the bias polarity, which changes the environment of the head group such, that the directed rotation does not work at the same z position and bias value. The fact that the directed rotation can be observed again, only by changing the tip position in z and the bias voltage, indicates a persistence of the directed rotation (within the limits of measurability).

Being able to measure TTs at the same lateral tip position at positive and negative bias brings checking for the inversion of the preferred direction within reach. The remaining problem is, that the preferred direction of rotation is only defined by increasing or decreasing the current. The single states cannot be imaged and it is unclear whether they represent a clockwise or counterclockwise rotation. Also, it is possible to observe that a current level with a higher current value decreases in current, upon changing z, until it drops below one of the other current levels. Therefore, on the two sides of this crossing point, the order of the levels with respect to the current is not the same. In example, let's assume that before the crossing point the order of states with increasing current is 0, 1, 2. Then, if state 2 drops in current below state 1, the order would be 0, 2, 1 after the crossing point. Therefore, it is not straightforward to find the corresponding states at positive and negative bias voltage.

However, to check for the inversion of the preferred rotation direction, it is only necessary to compare the rotation directions, the knowledge about the actual direction is not needed. Therefore, if the corresponding states at positive and negative bias can be correlated, it can be inferred, whether ascending transitions at positive bias correspond to ascending or descending transitions at negative bias. With this information, a possible inversion of the rotation direction can be detected.

To find the corresponding current levels at positive and negative bias, specific experiments were performed, called switching correspondences in the following. First, a lateral tip position had to be found, where TT measurements with good rejection probability were acquired, also defining the sets of bias voltage and z position for the two bias polarities, for which the correspondence between states has to be found.

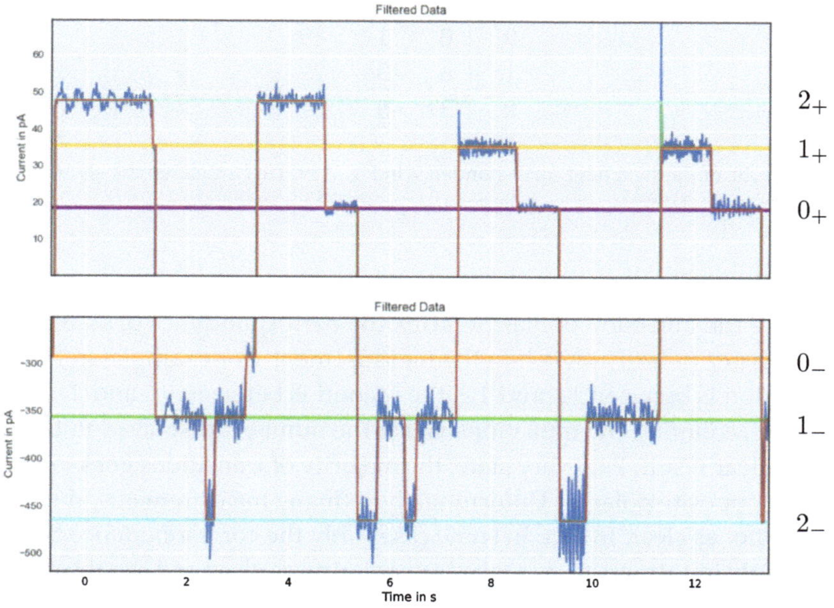

Figure 7.19: Part of a switching correspondence measurement. The measurement and time frame are the same, so once the signal gets out of range of one figure, it enters the other. Positive and negative bias are shown separately due to the difference in current range.

Using a *LabView*[3] program to control the measurement parameters, the z position and the bias voltage was switched periodically (usually every 2 s) between the two sets (usually for $200\,\text{s} \approx 3.3\,\text{min}$), while continuously measuring the current.
Part of a switching correspondence measurement is shown in Fig. 7.19. Because the currents are very different at positive and negative bias, the two current ranges are shown separately, but for the same time period of the same measurement, so once the signal gets out of range of one figure, it enters the other.
At both bias polarities, the current assumes three distinguishable levels and the switching between the polarities is fast compared to the residence time of the states. For the switching correspondence measurements, the three current levels are 0_+ to 2_+ at positive bias and 0_- to 2_- at negative bias, defined as shown in Fig. 7.19. The evaluation is done with an adjusted GUI, using the same evaluation methods as for TT measurements (compare Sec. 7.2).

[3] `https://www.ni.com/de-de/shop/labview.html`

	0_+	1_+	2_+
0_-	0	1	19
1_-	6	35	2
2_-	33	3	0

Table 7.1: Results of the switching correspondence in Fig. 7.19. This measurement shows a clear correspondence for every state.

Assuming that the head group does not switch into another metastable rotational state during the inversion of bias polarity, the correspondence of states can be deduced for every polarity switch in the measurement. For example, in Fig. 7.19, the first transition is between 1_+ and 1_-, the second is between 0_- and 2_+, etc. In this specific case, counting the transitions yields the numbers displayed in table 7.1 and has a very clear result. For every state, the majority of transitions goes to exactly one state at the opposite polarity. Unfortunately, in many measurements, the correspondences are not as clear. In fact, in some cases only the correspondence between two states (and their partners at opposite bias polarity) is clear and the third correspondence of states is inferred by exclusion.
In this example (Fig. 7.19, Tab. 7.1), an ascending transition at positive bias corresponds to an ascending transition at negative bias, called a non-inverting correspondence (ascending at positive bias corresponding to descending at negative bias is an inverting correspondence). The preferred direction for the TT measurements is ascending for both bias polarities, meaning that the rotation direction is not inverted upon inverting the bias polarity.
Despite the challenging experiment, six sets of measurements, including TTs at both bias polarities and the switching correspondence, were obtained. Their results are shown in Tab. 7.2. It should be noted, that the first three sets were measured on the same molecule and similar relative tip position. Also, it is assumed in the TT measurements, that two states can be seen as one, as the switching rate is very different. Moreover, the switching correspondences of the first 5 measurements do not yield obvious mapping of states at opposite polarities, either because one correspondence has to be inferred by exclusion or because the numbers are not as different for every single state mapping (such as $1_- \leftrightarrow 1_+$) as i.e. in Tab. 7.1.
To address this problem, the switching correspondence results can also be evaluated quantitatively. To test, whether the switching correspondence is inverting or not, a binomial test can be used, again yielding a rejection probability $p_{\text{reject, SC}}$. For that purpose, the sum of all transitions contributing to each set of state correspondences that is non-inverting (e.g. $2_- \leftrightarrow 0_+, 1_- \leftrightarrow 1_+, 0_- \leftrightarrow 2_+$) is computed. The set with the maximum number of transitions is taken to be $n_{\text{SC,non-inv}}$. The same is done for all inverting sets of state correspondences (e.g. $2_- \leftrightarrow 1_+, 1_- \leftrightarrow 0_+, 0_- \leftrightarrow 2_+$),

Bias (V)	p_{reject} (%)	Preferred Direction	Correspondence	Direction Inversion	$p_{\text{reject, SC}}$ (%)
1.75	> 99.99	Ascending	Inverting	No	97.36
−1.95	> 99.99	Descending			
1.7	> 99.99	Ascending	Non-Inverting	No	99.43
−1.9	> 99.99	Ascending			
1.9	99.70	Descending	Inverting	No	87.18
−1.9	99.47	Ascending			
1.9	99.98	Descending	Inverting	No	92.45
−1.73	> 99.99	Ascending			
1.87	> 99.99	Descending	Non-Inverting	Yes	75.84
−1.89	> 99.99	Ascending			
1.79	69.87	Ascending	Non-Inverting	No	99.99
−1.75	82.26	Ascending			

Table 7.2: Complete sets of TTs at both bias polarities and switching correspondences. The first three measurements are acquired on the same molecule. Most sets indicate that the rotation direction is not inverted upon inversion of the bias polarity.

giving $n_{\text{SC,inv}}$. This implies that the correspondence with the maximum number of transitions contributing (in the inverting and non-inverting case separately) is the most probable and is used for the binomial test. It is performed with $p_0 = 0.5$, $n_{\text{SC,inv}}$ and $n_{\text{SC,non-inv}}$ to test against the null hypothesis of both types of correspondences (inv. and non-inv.) being equally probable.

The results are shown in Tab. 7.2 and are good for most switching correspondences, except for two (one inverting the rotation direction and one not inverting it), indicating that the rotation direction is most probably not inverted.

The last measurement shown in Tab. 7.2 has a clear switching correspondence, but the rejection probabilities of the TTs are not very high.

In conclusion, most of the sets result in no inversion of the preferred direction of rotation for opposite bias polarity. As mentioned, this can give insight into the effect causing the directed motion. In a naïve picture this would exclude the chiral-induced spin selectivity effect, because the exerted torque by the CISS effect should point in the opposite direction, when injecting electrons on the opposite side of the chiral molecule. A more detailed description of possible effects and comparison with theoretical calculations is presented in Sec. 7.6.

7.5 Simulation of Tpdx-G

To get a better understanding of the underlying effect that drives the directed motion observed in Tpdx-G, *ab initio* calculations of Tpdx-G were performed by Dr. Jan Wilhelm in the group of Prof. Ferdinand Evers in Regensburg. The calculations include the adsorption geometry and molecular structure of Tpdx-G on a Au(111) surface (Subsec. 7.5.1) and the molecular orbitals of Tpdx-G (Subsec. 7.5.2).
The calculations help in verifying the chiral structure of the head group, in rationalizing the observed configuration changes of the molecule during rotation and give indications towards the driving mechanism. Finally in Sec. 7.6, different potential driving mechanisms of the rotation are reviewed concerning their compatibility with the experimental observations presented in Chap. 7.

7.5.1 Structure Simulation

The first step to gaining more insight into the cause of the observed directed rotation is the verification of the molecular structure of Tpdx-G and its adsorption on the Au(111) surface. For that purpose, density functional theory (DFT) calculations are performed, relaxing the molecular structure on a four monolayer thick slab of the Au(111) surface, including Van der Waals interactions.
The resulting molecular structure (without the Au surface) is shown in Fig. 7.20 at two different angles. The view in Fig. 7.20a is onto the two benzene rings connected by the Geländer structure. The teal, white, yellow and red lines represent bonds between carbon, hydrogen, sulfur and oxygen atoms, respectively.
From this view, it is apparent, that the molecule does not have a tetragonal shape, as might be expected from the molecular structure, but the foot structures are arranged in a plane, due to attractive interaction with the surface. Moreover, it is visible, that the head group does not point perpendicularly away from the surface or the plane of the foot structures, but encloses an angle of $\sim 20\,^\circ$ with the perpendicular.
As briefly discussed in Subsec. 7.4.2, the observed variation in tip–sample distance is of the order of few $10\,\mathrm{pm}$ to $100\,\mathrm{pm}$, estimated from the current differences in TT measurements. Measuring the lengths in the relaxed molecular structure, a rotation of the complete head group around the perpendicular through the central carbon atom (connecting the three feet and the head group), would yield a much too large movement with a radius of $\sim 380\,\mathrm{pm}$ (Fig. 7.21b). Nevertheless, such a movement, together with the superimposed shape of the STM tip, can explain the shape of Tpdx-G molecules observed in constant current images. The large radius of the molecule, together with the indications of abrupt movement during scanning are compatible with a larger-scale movement of the head group.

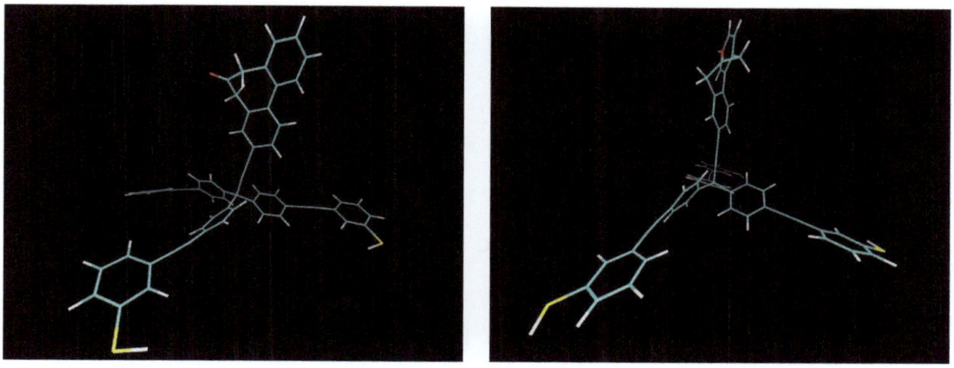

(a) View perpendicular to the plane of the head group benzene rings.

(b) View parallel to the plane of the head group benzene rings.

Figure 7.20: Two views on the relaxed molecular structure, not including the Au surface slab. The teal, white, yellow and red lines represent bonds between carbon, hydrogen, sulfur and oxygen atoms, respectively. The head structure does not point exactly perpendicular to the foot structure plane, but is at an angle. It also has a tilt in the planes of the two benzene rings. The foot structures are rather flat and not pyramidal.

However, another, smaller rotation is possible, because the head group also has an angle in itself. Measuring the angle between the central carbon atom, the first and the last carbon atom of the head group (Fig. 7.21a), yields a radius of rotation of 72 pm around the C–C bond connecting the head group to the central carbon atom. A movement on this length scale is much more probable, as it is compatible with the current differences in TT measurements. The rotation being on such a small length scale can also explain, that the metastable rotational states cannot be imaged, as transitions between such conformations are easily induced by the tip movement during scanning.
In Fig. 7.20b, the molecule is shown from a view parallel to the benzene rings in the head group, depicting the conformation of the head group more clearly. It shows, that the planes of the two benzene rings of the head group are at an angle, meaning, that the Geländer structure actually serves its purpose. It tilts the benzene rings in such a way, that the structure becomes chiral.
Knowing the adsorption geometry of the molecule, the molecular orbitals can be calculated.

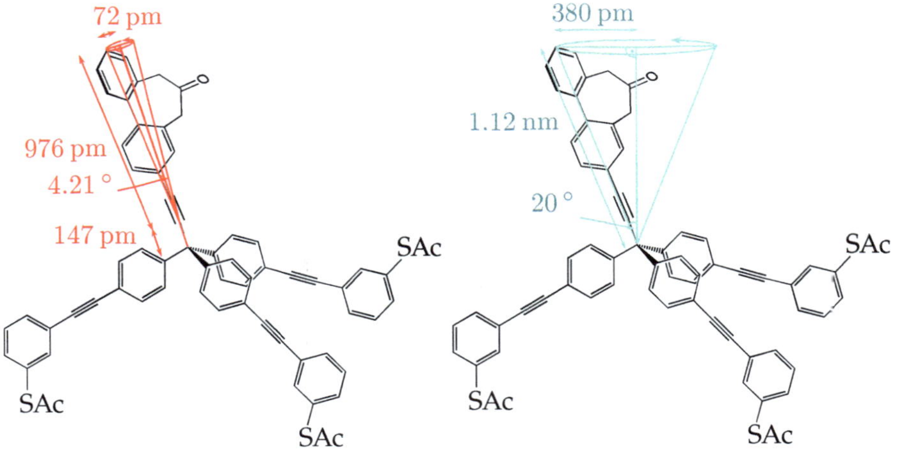

(a) Sketch of the smaller rotation.

(b) Sketch of the larger rotation.

Figure 7.21: Chemical structure with lengths extracted from the simulated structure. A rotation around the backbone of the Geländer group (red) yields length scales of an order compatible with the current changes in TT measurements. A rotation around the perpendicular onto the central carbon atom (teal) is too large.

7.5.2 Simulated Molecular Orbitals

It is seen in the bias voltage dependence, that the directed motion can only be observed at specific bias voltages, not being observable at higher or lower voltages, indicating that tunneling into a specific molecular orbital (MO) plays a crucial role. The intention of the simulation is to find characteristic features in the MOs that explain the specific energies, at which directed rotation is observed.

Using the molecular geometry from Subsec. 7.5.1 in gas phase, the MOs are calculated using a self-consistent simulation in the GW_0 approximation using the Perdew-Burke-Ernzerhof (PBE) functional. Exemplary MOs are shown in Fig. 7.22. The calculations yield the orbital function $\psi_{\mathrm{MO}}(x,y,z)$ for the MO at a certain energy. To display a MO sensibly, it is plotted as a so-called isosurface, which is the surface of the orbital function at a certain isovalue $\psi_{\mathrm{MO}}(x,y,z) = \psi_{\mathrm{MO,Iso}}$. The red and blue color of the surfaces represent positive and negative isovalues, respectively.

The isovalue has to be chosen appropriately to display the MO representatively, as the orbital function decays with the distance from the atoms. If the isovalue is too small, the isosurface is very large and far away from the atoms and if the isovalue is too large, the isosurface is very small, not being displayed even though $\psi_{\mathrm{MO}}(x,y,z)$ has reasonable values. In both cases, the MO is not represented properly. An appropriate value in this case is $\psi_{\mathrm{MO,Iso}} = 0.015$. Since the electron density is proportional to the absolute square of the wave function, it can also be seen as the electron density of the corresponding MO.

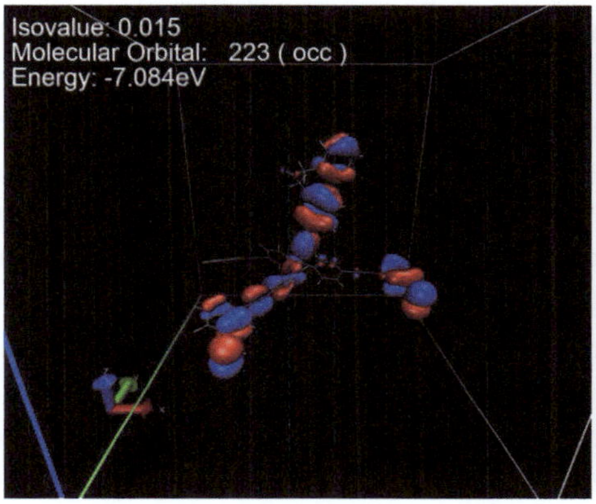

(a) Occupied MO with wave function only on the backbone of the head group.

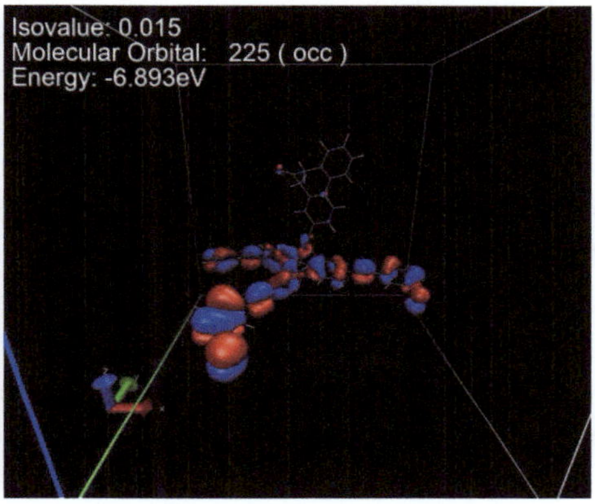

(b) Occupied MO with little wave function amplitude on the head group.

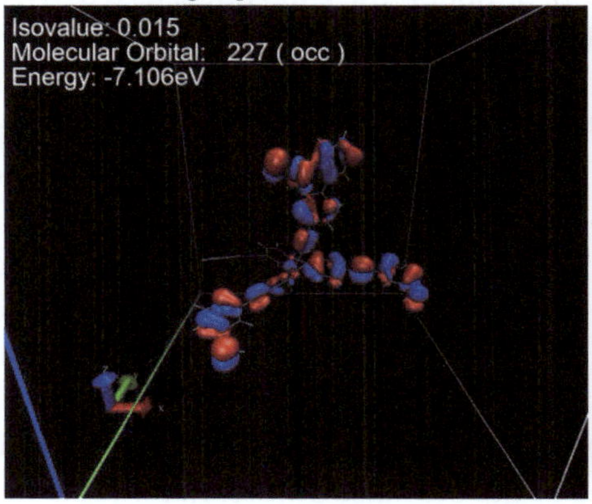

(c) Occupied MO with significant wave function amplitude on both backbone and Geländer.

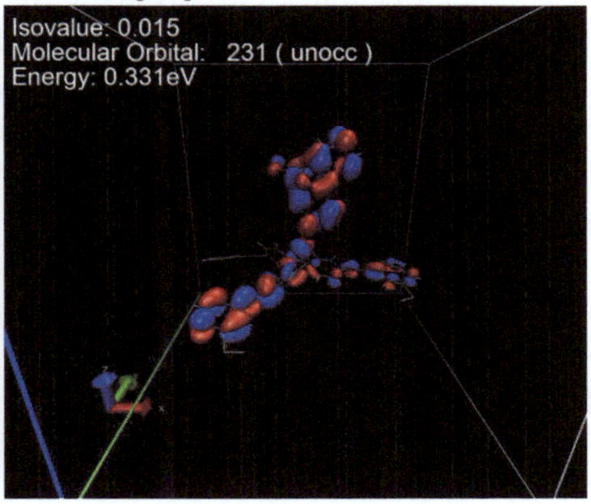

(d) Unoccupied MO with significant wave function amplitude on both backbone and Geländer.

Figure 7.22: Calculated molecular orbitals. All representations are shown from the same angle as Fig. 7.20a. They show various characteristics, going from having almost no wave function amplitude on the head group, to having it only on the backbone to having it on both the backbone and the Geländer.

The different MOs have very different electron densities distributed over the molecule, as shown in Fig. 7.22. In Fig. 7.22a, for example, there is a distribution that contributes to a current running through the molecule, as the electron density forms a pathway from the head group to the foot structure and therefore into the substrate. However, the electron distribution on the head group is only localized on the backbone and not on the Geländer, which connects the benzene rings of the head group.
Figure 7.22b shows a MO that does not have any electron density on the head group, thus not allowing for any current to be carried by this molecular orbital. The MOs depicted in Fig. 7.22c and Fig. 7.22d are very similar, but are at very different energies and are occupied and unoccupied orbitals, respectively.
Some characteristics could influence whether the MO contributes to driving the directed motion, i.e. by defining, if the prerequisites for CISS are met. One such criterion could be, if there is electron density (and therefore charge density) on the Geländer or not, which, in the simple picture of CISS (compare Subsec. 2.4.1), would correspond to the helical electron distribution.
In fact, there are both occupied and unoccupied MOs with electron density on the Geländer group, surrounded energetically by either MOs without electron density on the Geländer or the entire head group. This observation is in agreement with the conceptual idea that CISS drives the rotation, although the CISS effect as a drive is excluded due to other observations, as elaborated in Sec. 7.6.
Overall, the molecular orbitals are very similar in their shape, having the same succession of positive and negative valued $\psi_{\mathrm{MO}}(x,y,z)$ throughout the head group and similar shapes of the MOs (compare Fig. 7.22c and 7.22d). Only the size of the isosurfaces at the different sites varies between MOs. However, they are still located closely to the atoms and e.g. do not change sign between occupied and unoccupied MOs. As the electrons tunnel into unoccupied orbitals at positive bias and occupied orbitals at negative bias, this means that there is no qualitative difference between the shapes of the MOs depending solely the bias polarity.
To compare the simulated MOs to experimental data, i.e. from the voltage dependence, it is crucial to consider the energies of the orbitals. Unfortunately, it is not possible to extract the relative position of the Fermi energy with respect to the MOs, which means that only energetic differences between MOs can be interpreted meaningfully.
Taking a closer look at these energies (indicated in Fig. 7.22), it becomes clear that they are very large. The difference between the highest occupied MO (HOMO) and the lowest unoccupied MO (LUMO), which is comparable to the energy difference between current-carrying occupied and unoccupied MOs, is $\sim 7\,\mathrm{eV}$, corresponding to a gap in the spectrum of 7 V. However, as mentioned before, the calculations of the MOs are done for the gas-phase molecule.

Including the Au surface introduces screening of the charge density. In the calculations, the screening is modeled by adding an image charge on the opposite side of the Au surface and renormalizing the MO energies by the Coulomb interaction between charge and image charge [89]. For reasonable computation times and simplification of the calculation, the orbitals used in this process are restricted to the π-orbitals of the foot structures.
This renormalization yields a significant reduction of the HOMO–LUMO gap to about $3 - 4$ eV. Comparing this energy difference with the experimental voltages at which sets of measurements at both polarities could be obtained (see Tab. 7.2), they yield compatible voltage differences of 3.67 V on average. Averaging over all TT measurements with high p_{reject} gives a voltage difference of 3.72 V, which is also compatible.
Lastly, calculations of energy barrier heights between metastable rotational states similar to the sketch shown in Fig. 7.21 were performed. The difference to the rotation described therein is that the head group rotates around the axis defined by its two triple-bonded carbon atoms, not around the C–C bond connecting the head group with the central carbon atom. The simulation results are shown in Fig. 7.23 and present a potential landscape with two minima separated by a $180\,^{\circ}$ rotation angle and a maximum barrier height of 12 meV.
For a rough estimation of the barrier heights in the experiment, a simple Arrhenius law is used

$$k_{\text{trans}} = \nu_{\text{try}} \exp \frac{E_{\text{trans}}}{k_B T}\,, \tag{7.8}$$

where E_{trans} is the transition barrier height, $\nu_{\text{try}} = 10^{11}\,\frac{1}{\text{s}}$ is the frequency of trying to cross the barrier and k_{trans} is the rate constant of the transition. All quantities needed to calculate the barrier energy E_{trans} are known or taken from experiments (k_{trans} is assumed to be the switching rate of the corresponding transition), resulting in energy barriers of $\sim 9 - 12$ meV, in very good agreement with the simulation.
A disagreement between the simulation and the experiment is the number of minima in the potential landscape. For the directed motion, three distinct metastable current levels are observed, whereas the simulation only exhibits two minima. An explanation for this discrepancy could be the presence of the tip, which is not included in the simulation, but can have significant effects on the potential landscape, possibly introducing a third minimum.
On the basis of both the experimental data and theoretical calculations of the molecular structure and orbitals, different causes of the directed rotation are evaluated in Sec. 7.6.

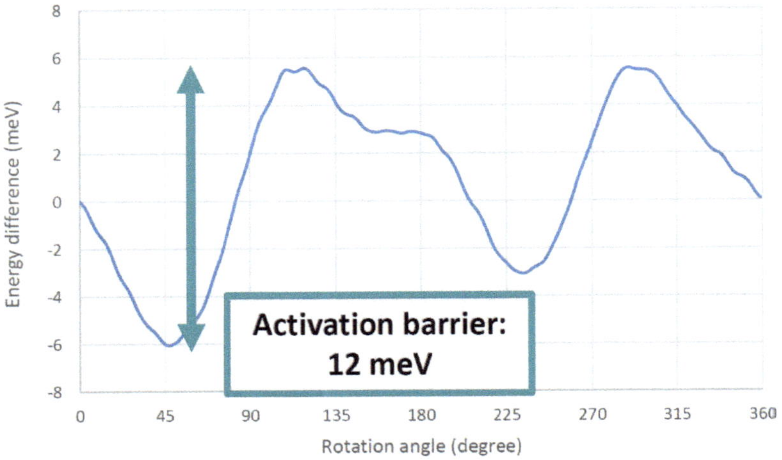

Figure 7.23: Simulated barriers of a rotation around the triple bond of the head group. Although it exhibits only two minima, the barrier heights match estimations from the observed rates, based on the Arrhenius law, for a similar rotation (red in Fig. 7.21) (image provided by Dr. Jan Wilhelm).

7.6 Compatibility of Observed Rotational Behavior of Tpdx-G with Potential Drives

With the experimental observations discussed in detail in Sec. 7.4 and the theoretical calculations presented in Sec. 7.5, the proposed drives (see Sec. 2.4) can be tested. The intended mechanism to drive the directed rotation is the chiral-induced spin selectivity (CISS) effect. As discussed in more detail in Subsec. 2.4.1, the spin-polarizing effect of the chiral molecule is supposed to drive the rotation *via* spin-orbit coupling. This effect thus depends on the chirality of the molecule and the electron density as described by the molecular orbitals (MOs). As mentioned in Subsec. 7.5.2, MOs with electron density on the Geländer group can be found at energy differences compatible with experimental data from the switching correspondences.
However, the CISS effect can most likely be excluded as the driving mechanism, since the switching correspondences indicate, that the preferred rotation direction is not inverted for opposite bias polarities. CISS as the drive would rely on the spin-filtering properties of the chiral molecules and should therefore lead to inverted preferred rotation directions for opposite bias polarities.
An instructional example is a helical structure, through which a current is run in opposite directions, as depicted in Fig. 7.24. As the spin-polarization effect depends on the direction of movement of the electrons carrying the current, the filtered spin is either parallel or antiparallel to the direction of the current. Assuming the spins antiparallel to the current direction are filtered and the parallel ones can pass the

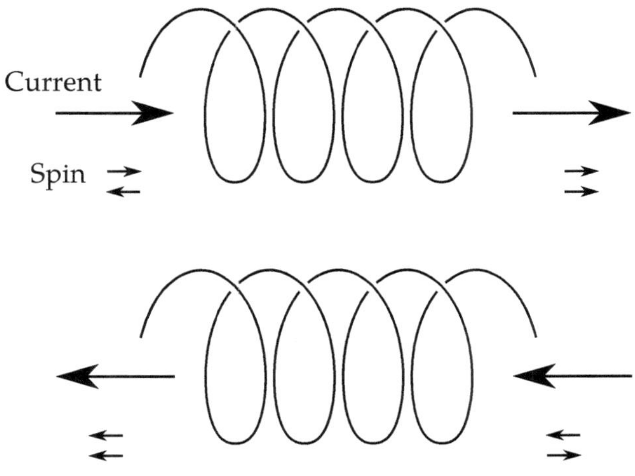

Figure 7.24: Sketch of the spin-polarization caused by CISS for different current directions. The helix direction is the same with respect to the current direction. The spin-filtering ability only depends on their relative alignment. Therefore, the spins are filtered in opposite directions for inverted current direction.

molecule, the opposite spin-polarity is filtered, if the current direction is inverted. This holds also for the assumption that parallel spins are filtered and antiparallel spins can pass the molecule.
For the specific case of the rotating head group, this means, that the direction of the back action angular momentum is also inverted for opposite current directions, in turn inverting the preferred direction of rotation. However, as presented in Subsec. 7.4.4, the rotation direction is not inverted for opposite bias polarities and therefore opposite current directions.
It can also be excluded that the MOs carrying the current at positive and negative bias are so different, that the spin-polarization properties of the head group are inverted, as the occupied and unoccupied MOs are qualitatively very similar. This means, that CISS is very unlikely to serve as the drive of the observed directed rotation of the Tpdx-G head group. In principle, the same argument holds for all processes that filter spins depending on the direction of current, e.g. processes exploiting spin-orbit coupling.
The other proposed drive for directed rotation of the head group, presented in Subsec. 2.4.2, is the on-off ratchet. For this mechanism to work, an asymmetric potential and repetitively lowered and raised potential barriers between the metastable states are needed.
As shown in the theoretical simulation of the potential barriers for a rotation of the head group (Fig. 7.23), an asymmetric potential is probable, but no varying bias voltage (or other stimulus), which could repetitively raise and lower the barrier heights, is actively applied. Another possible cause for the variation in barrier heights are

thermal or statistical fluctuations (such as shot noise). However, these fluctuations cannot perform harvestable work and since a directed rotation is a process that inherently performs harvestable work, it cannot be caused by the aforementioned fluctuations. This means that it is very unlikely that the directed rotation is driven by the on-off ratchet effect.

Finally, inelastically tunneling electrons can deposit energy in a system, i.e. as vibrational energy, which could be transformed into directed rotation. As discussed briefly in Subsec. 2.1.2, a new conduction channel opens at a threshold electron energy, or bias voltage, increasing the slope in a I/U spectrum. However, this is not visible in Fig. 7.17a. Generally, if the directed rotation is driven by inelastic excitation, its contribution to the current and its rate of excitation should increase with increasing bias voltage (see Subsec. 7.4.3), as more electrons have the energy to excite it. As the asymmetry (and velocity) show, this is not the case for the directed rotation of Tpdx-G, as they decrease with increasing bias voltage.

Summing up, it remains an open question which mechanism is responsible for the directed rotation observed in Tpdx-G molecules and it might well be possible that it is a combination of more than one. To the best of our knowledge, with the experimental data and the simulations presented in this thesis, no clear conclusion is possible, but some effects, such as CISS and the on-off ratchet can probably be excluded.

8 Conclusion and Outlook

The aim of this thesis is to find and study the directed rotation of a single, self-decoupled molecule which works as an electronic motor, driven by the tunneling current in an STM. For that purpose, different tripodal molecules are investigated, finding a way to reproducibly isolate the rotator of a single molecule, allowing to address it with the STM tip. A directed rotation was indeed observed and the high reproducibility allows for a detailed statistical analysis of this directed rotation. Dependences on various parameters were studied in an attempt to grasp the underlying effect driving the rotation. However, in combination with theoretical simulations, the cause cannot be identified unambiguously and the rotation might well be caused by multiple effects. As the chiral-induced spin selectivity (CISS) effect was considered as a candidate to cause the directionality, it is also tried to study the effect itself, using conductive atomic force microscopy (CAFM). This chapter briefly sums up the steps taken from the search for suitable molecules to the interpretation of characteristics of a directed rotation of a nanomotor.

The first step is to find a molecule that allows for the study of directed rotation. The path chosen here is to use tripodal platforms with a mounted head group, which is supposed to be freely rotatable and transfer electric current into a directed motion. The first investigated molecule uses a smaller type of platform with a mounted chiral rotator (Tpd-G). These molecules, however appear to adsorb on the surface with the head group, rather than the platform. Even increasing the rigidity of the platform only allows the molecules to have the head group point away from the surface, if they are not embedded in an ordered island (sTpd-L). Both molecules are therefore not suitable to study directed rotation of the head group.
Therefore, the platform was extended and various head groups are mounted, namely a hydrogen atom (Tpdx-H), another foot structure (Tpdx-S), a chiral rotator (Tpdx-G) and an extended rotator with an additional benzene ring (Tpdx-Gx).
Depositing the first three molecules yields similar ordered arrangements of the molecules, indicating that they adsorb in the same way. As Tpdx-H has only a hydrogen atom as a head group, the molecules adsorb with the tripodal platform on the surface. For single Tpdx-G molecules, it is possible to observe directed rotation. However, the metastable states of the rotation cannot be imaged and the directed

rotation is only observed in specific cases. Adding a benzene ring to the head group (Tpdx-Gx) is intended to make imaging of the metastable states possible, but leads to the molecule adsorbing with the head group on the surface.
To achieve more reproducible observation of directed rotation, Tpdx-G molecules were deposited together with Tpdx-H and form intermixed islands, where single Tpdx-G molecules are embedded in a lattice of Tpdx-H. This provides a more stable environment for Tpdx-G molecules and indeed allows for a reliable observation of directed rotation, which is necessary as a large database is needed to statistically analyze the directed rotation.

With a system that allows for reliable observation of directed motion, it can be studied in more detail. The metastable states cannot be imaged directly, but the tunneling current shows switching between three distinguishable current levels, which are interpreted as metastable rotational states that are assumed during the rotation. As three states are observable, a direction can be defined, depending on whether the current is increased or decreased for a switching event.
As many measurements have to be performed, a graphical user interface is programmed to efficiently evaluate measurements of the tunneling current over time, called time traces (TTs), and extract the number of transitions between the three current levels. Whether the observed rotation is directed or not, is determined by performing a binomial test with the counted number of transitions in one direction and the other, revealing that many measurements exhibit directed rotation with a certainty close to 100 %.
To extract physical quantities of the rotation, every switching event is interpreted as a rotation of the head group by $\pm 120\,^{\circ}$, depending on the direction. This allows for plots of the angle covered by the head group over time. These plots turn out to depend linearly on time, indicating dominant directed rotation and enabling the calculation of the velocity of the rotation through a linear fit. To verify this method, the directed rotation is also evaluated by means of the mean square displacement, showing good agreement.
For a study of the directed rotation and the underlying effect driving it, its dependence on the lateral tip position, bias voltage, tunneling current and bias polarity is observed. As a first attempt, figures of merit, such as asymmetry and velocity, are represented in scatter plots, showing that their variation is large compared to the parameter range, in which TT measurements can be acquired.
As the measurements included in the scatter plots are performed across various molecules and relative tip–molecule positions, the dependence on the lateral tip position is investigated as a candidate causing the large variation. This was done by measuring grids of TTs around a lateral tip position for which directed rotation is observed. They reveal that the directed rotation depends strongly on the lateral tip

position, as it is only observable on a length scale of $\sim 100\,\mathrm{pm}$. This also explains, why the metastable rotational states are not observable in STM images. Due to the strong lateral tip position dependence, conclusions can only be drawn from measurement series performed for a fixed lateral tip position.
Current dependent measurements also show a strong influence on the directed rotation, as the tip position has to be adjusted to change the current. Counterintuitively, the higher currents lead to lower switching rates. In voltage dependent measurements, non-monotonic behavior is observed for the asymmetry, indicating tunneling into a specific molecular orbital to drive the directed rotation. Moreover, TTs at positive and negative bias voltages are compared, revealing that the direction of rotation is not inverted for opposite bias polarities.
To gain insight into the underlying effect, simulations of the molecular structure and orbitals are performed in the group of Prof. F. Evers. They show that the occupied and unoccupied molecular orbitals are not qualitatively different. Theoretical calculations of barrier heights for a rotation, similar to the observed one, agree with estimations from experimental data.
Combining the experimental observations and theoretical simulations, potential effects driving the directed motion, such as CISS, the on-off ratchet effect and inelastic tunneling, can most likely be excluded. It therefore remains an open question which effect drives the directed rotation and it is possible that not only one effect is responsible.

In supporting experiments performed at Leiden University, it was tried to study the CISS effect directly, by measuring the spin-dependent conductance through self-assemled monolayers of helical molecules using conductive atomic force microscopy. However, reproducibility of these measurements suffers from a number of varying parameters, so that only qualitative observations regarding the choice of substrate for comparable current measurements on molecular self-assembled monolayers can be made.

For a further investigation of the effect causing directed rotation in Tpdx-G, the molecular structure could be adjusted, e.g. by straightening the head group. That way, the influence of the helicity could be excluded and it could be tested, if the rotation can still be observed and if it is still directed, providing experimental proof, whether CISS plays a role in causing the observed rotation.
Moreover, the studies presented in this thesis present a modular platform, which can be used to self-decouple functional groups from the surface in a reliable way, potentially in a codeposition with Tpdx-H, which can be expanded to other functionalities, such as light-emitting molecules.

Bibliography

[1] *Nobel Prize Web Page on Bernard Feringa*. URL: https://www.nobelprize.org/prizes/chemistry/2016/feringa/facts/.

[2] N. Koumura, R. W. Zijistra, R. A. Van Delden, N. Harada, and B. L. Feringa. "Light-driven monodirectional molecular rotor". In: *Nature* 401 (1999), pp. 152–155. DOI: 10.1038/43646.

[3] J. Bardeen. "Tunneling from a many-particle point of view". In: *Physical Review Letters* 6 (1961), pp. 57–59. DOI: 10.1103/PhysRevLett.41.1825.

[4] D. R. Hamann and J. Tersoff. "Theory of the scanning tunneling microscope". In: *Phsical Review B* 31 (1985), pp. 805–813. DOI: doi.org/10.1103/PhysRevB.31.805.

[5] B. C. Stipe, M. A. Rezaei, and W. Ho. "Single-molecule vibrational spectroscopy and microscopy". In: *Science* 280 (1998), pp. 1732–1735. DOI: 10.1126/science.280.5370.1732.

[6] H. Gawronski, M. Mehlhorn, and K. Morgenstern. "Imaging phonon excitation with atomic resolution". In: *Science* 319 (2008), pp. 930–933. DOI: 10.1126/science.1152473.

[7] A. J. Heinrich, J. A. Gupta, C. P. Lutz, and D. M. Eigler. "Single-atom spin-flip spectroscopy". In: *Science* 306 (2004), pp. 466–470.

[8] J. P. Gauyacq, N. Lorente, and F. D. Novaes. "Excitation of local magnetic moments by tunneling electrons". In: *Progress in Surface Science* 87 (2012), pp. 63–107. DOI: 10.1016/j.progsurf.2012.05.003.

[9] G. Binning, C. F. Quate, and C. Gerber. "Atomic force microscope". In: *Physical Review Letters* 56 (1986), pp. 930–933. DOI: 10.1201/9781420075250.

[10] TimeStep89. *Lennard-Jones potential wikipage*. URL: https://en.wikipedia.org/wiki/Lennard-Jones_potential.

[11] S. Tien, S. Devasia, and Q. Zou. "Iterative control of dynamics-coupling-caused errors in piezoscanners during high-speed AFM operation". In: *IEEE Transactions on Control Systems Technology* 13 (2005), pp. 921–931. DOI: 10.1109/TCST.2005.854334.

[12] T. Göddenhenrich, H. Lemke, U. Hartmann, and C. Heiden. "Force microscope with capacitive displacement detection". In: *Journal of Vacuum Science and Technology A: Vacuum, Surfaces, and Films* 8 (1990), pp. 383–387. DOI: 10.1116/1.576401.

[13] F. J. Giessibl. "High-speed force sensor for force microscopy and profilometry utilizing a quartz tuning fork". In: *Applied Physics Letters* 73 (1998), pp. 3956–3958. DOI: 10.1063/1.122948.

[14] J. E. Sader, J. W. Chon, and P. Mulvaney. "Calibration of rectangular atomic force microscope cantilevers". In: *Review of Scientific Instruments* 70 (1999), pp. 3967–3969. DOI: 10.1063/1.1150021.

[15] J. E. Sader. "Frequency response of cantilever beams immersed in viscous fluids with applications to the atomic force microscope John". In: *Journal of Applied Physics* 84 (1998), pp. 64–76. DOI: 10.1002/9783527676330.ch2.

[16] N. A. Burnham, X. Chen, C. S. Hodges, G. A. Matei, E. J. Thoreson, C. J. Roberts, M. C. Davies, and S. J. Tendler. "Comparison of calibration methods for atomic-force microscopy cantilevers". In: *Nanotechnology* 14 (2003), pp. 1–6. DOI: 10.1088/0957-4484/14/1/301.

[17] *Park Systems Phase imaging*. URL: https://www.parksystems.com/park-spm-modes/91-standard-imaging-mode/221-phase-imaging-phase-detection-microscopy-pdm.

[18] H. Hertz. "Über die Berührung fester elastischer Körper". In: *Journal für die reine und angewandte Mathematik* 171 (1881), pp. 156–171.

[19] I. N. Sneddon. "The relation between load and penetration in the axisymmetric boussinesq problem for a punch of arbitrary profile". In: *International Journal of Engineering Science* 3 (1965), pp. 47–57. DOI: 10.1016/0020-7225(65)90019-4.

[20] G. G. Bilodeau. "Regular pyramid punch problem". In: *Journal of Applied Mechanics, Transactions ASME* 59 (1992), pp. 519–523. DOI: 10.1115/1.2893754.

[21] P. Gruetter, H. J. Mamin, and D. Rugar. "Magnetic Force Microscopy (MFM)". In: *Scanning Tunneling Microscopy II: Further Applications and Related Scanning*

Techniques. Berlin, Heidelberg: Springer Berlin Heidelberg, 1992, pp. 151–207. DOI: 10.1007/978-3-642-97363-5_5.

[22] R. D. Gomez, E. R. Burke, and L. D. Mayergoyz. "Magnetic imaging in the presence of external fields: Technique and applications (invited)". In: *Journal of Applied Physics* 79 (1996), pp. 6441–6446. DOI: 10.1063/1.361966.

[23] H. Hölscher, U. D. Schwarz, and R. Wiesendanger. "Calculation of the frequency shift in dynamic force microscopy". In: *Applied Surface Science* 140 (1999), pp. 344–351. DOI: 10.1016/S0169-4332(98)00552-2.

[24] T. Häberle, F. Haering, H. Pfeifer, L. Han, B. Kuerbanjiang, U. Wiedwald, U. Herr, and B. Koslowski. "Towards quantitative magnetic force microscopy: Theory and experiment". In: *New Journal of Physics* 14 (2012), p. 043044. DOI: 10.1088/1367-2630/14/4/043044.

[25] F. J. Giessibl. "Forces and frequency shifts in atomic-resolution dynamic-force microscopy". In: *Physical Review B - Condensed Matter and Materials Physics* 56 (1997), pp. 16010–16015. DOI: 10.1103/PhysRevB.56.16010.

[26] R. D. Vale, T. S. Reese, and M. P. Sheetz. "Identification of a novel force-generating protein, kinesin, involved in microtubule-based motility". In: *Cell* 42 (1985), pp. 39–50. DOI: 10.1016/S0092-8674(85)80099-4.

[27] J. A. Spudich. "The myosin swinging cross-bridgemodel". In: *Nature Reviews Molecular Cell Biology* 2 (2001), pp. 387–392.

[28] S. Kojima and D. F. Blair. "The Bacterial Flagellar Motor: Structure and Function of a Complex Molecular Machine". In: *International Review of Cytology: A Survey of Cell Biology*. Academic Press, 2004. DOI: https://doi.org/10.1016/S0074-7696(04)33003-2.

[29] C. Richardson. *The flagellar hook: Making sense of bacterial motility*. 2019. URL: https://www.oist.jp/news-center/press-releases/flagellar-hook-making-sense-bacterial-motility.

[30] T. R. Kelly, H. De Silva, and R. A. Silva. "Unidirectional rotary motion in a molecular system". In: *Nature* 401 (1999), pp. 150–152. DOI: 10.1038/43639.

[31] H. L. Tierney, C. J. Murphy, and E. C. H. Sykes. "Regular scanning tunneling microscope tips can be intrinsically chiral". In: *Physical Review Letters* 106 (2011), p. 010801. DOI: 10.1103/PhysRevLett.106.010801.

[32] S. Stolz, O. Gröning, J. Prinz, H. Brune, and R. Widmer. "Molecular motor crossing the frontier of classical to quantum tunneling motion". In: *Proceedings*

of the National Academy of Sciences of the United States of America 117 (2020), pp. 14838–14842. DOI: 10.1073/pnas.1918654117.

[33] U. G. Perera, F. Ample, H. Kersell, Y. Zhang, G. Vives, J. Echeverria, M. Grisolia, G. Rapenne, C. Joachim, and S. W. Hla. "Controlled clockwise and anticlockwise rotational switching of a molecular motor". In: *Nature Nanotechnology* 8.1 (2013), pp. 46–51. DOI: 10.1038/nnano.2012.218.

[34] O. W. Richardson. "A mechanical effect accompanying magnetization". In: *Physical Review* 26 (1908), pp. 248–253. DOI: 10.1103/PhysRevSeriesI.26.248.

[35] A. Einstein and W. J. de Haas. "Experimental proof of the existence of Ampère's molecular currents." In: *KNAW, Proceedings, 18 I.* Amsterdam, 1915.

[36] T. Ono and H. Kohno. "Spin-transfer motor". In: *Journal of the Magnetics Society of Japan* 31 (2007), p. 305.

[37] R. Naaman and D. H. Waldeck. "Chiral-induced spin selectivity effect". In: *Journal of Physical Chemistry Letters* 3 (2012), pp. 2178–2187. DOI: 10.1021/jz300793y.

[38] E. Medina, F. López, M. A. Ratner, and V. Mujica. "Chiral molecular films as electron polarizers and polarization modulators". In: *Europhysics Letters* 99 (2012), p. 17006. DOI: 10.1209/0295-5075/99/17006.

[39] R. Gutierrez, E. Díaz, R. Naaman, and G. Cuniberti. "Spin-selective transport through helical molecular systems". In: *Physical Review B - Condensed Matter and Materials Physics* 85 (2012), p. 081404. DOI: 10.1103/PhysRevB.85.081404.

[40] K. Ray, S. P. Ananthavel, D. H. Waldeck, and R. Naaman. "Asymmetric scattering of polarized electrons by organized organic films of chiral molecules". In: *Science* 283 (1999), pp. 814–816. DOI: 10.1126/science.283.5403.814.

[41] V. Kiran, S. R. Cohen, and R. Naaman. "Structure dependent spin selectivity in electron transport through oligopeptides". In: *Journal of Chemical Physics* 146 (2017), p. 092302. DOI: 10.1063/1.4966237.

[42] M. Suda, Y. Thathong, V. Promarak, H. Kojima, M. Nakamura, T. Shiraogawa, M. Ehara, and H. M. Yamamoto. "Light-driven molecular switch for reconfigurable spin filters". In: *Nature Communications* 10 (2019), p. 2455. DOI: 10.1038/s41467-019-10423-6.

[43] G. L. Rikken and E. Raupach. "Enantioselective magnetochiral photochemistry". In: *Nature* 405 (2000), pp. 932–935. DOI: 10.1038/35016043.

[44] F. Evers, A. Aharony, N. Bar-Gill, O. Entin-Wohlman, P. Hedegård, O. Hod, P. Jelinek, G. Kamieniarz, M. Lemeshko, K. Michaeli, V. Mujica, R. Naaman, Y. Paltiel, S. Refaely-Abramson, O. Tal, J. Thijssen, M. Thoss, J. M. van Ruitenbeek, L. Venkataraman, D. H. Waldeck, B. Yan, and L. Kronik. "Theory of chirality induced spin selectivity: Progress and challenges". In: *Advanced Materials* 34 (2022), p. 2106629. DOI: 10.1002/adma.202106629.

[45] F. Evers and J. M. van Ruitenbeek. *Private Communication.*

[46] S. N. Ethier and J. Lee. "The flashing brownian ratchet and parrondo's paradox". In: *Royal Society Open Science* 5 (2018), p. 171685. DOI: 10.1098/rsos.171685.

[47] A. L. R. Bug and B. J. Berne. "Shaking-Induced Transition to a Nonequilibrium State". In: *Physical Review Letters* 59 (1987), p. 948.

[48] A. Ajdari and J. Prost. "Mouvement induit par un potentiel périodique de basse symétrie: Diélectrophorèse pulsée". In: *Comptes rendus de l'Académie des sciences* 315 (1992), pp. 1635–1639.

[49] S. Litvinov, M. Ellero, X. Hu, and N. A. Adams. "Self-diffusion coefficient in smoothed dissipative particle dynamics". In: *Journal of Chemical Physics* 130 (2009), p. 021101. DOI: 10.1063/1.3058437.

[50] M. Baum, F. Erdel, M. Wachsmuth, and K. Rippe. "Retrieving the intracellular topology from multi-scale protein mobility mapping in living cells". In: *Nature Communications* 5 (2014), p. 4494. DOI: 10.1038/ncomms5494.

[51] V. Tejedor, O. Bénichou, R. Voituriez, R. Jungmann, F. Simmel, C. Selhuber-Unkel, L. B. Oddershede, and R. Metzler. "Quantitative analysis of single particle trajectories: Mean maximal excursion method". In: *Biophysical Journal* 98 (2010), pp. 1364–1372. DOI: 10.1016/j.bpj.2009.12.4282.

[52] G. Ehrlich and F. G. Hudda. "Atomic view of surface self-diffusion: Tungsten on tungsten". In: *The Journal of Chemical Physics* 44 (1966), pp. 1039–1049. DOI: 10.1063/1.1726787.

[53] J. D. Doll and A. F. Voter. "Recent Developments in the Theory of Surface Diffusion". In: *Annual Review of Physical Chemistry* 38 (1987), pp. 413–431.

[54] R. Metzler and J. Klafter. "The restaurant at the end of the random walk: Recent developments in the description of anomalous transport by fractional dynamics". In: *Journal of Physics A: Mathematical and General* 37 (2004), R161–R208. DOI: 10.1088/0305-4470/37/31/R01.

[55] L. E. Reichl. *A modern course in statistical physics / by L.E. Reichl.* 2nd ed. New York: Wiley, 1998.

[56] Jkrieger. *Mean square displacement wikipage.* URL: https://de.wikipedia.org/wiki/Mittlere_quadratische_Verschiebung.

[57] Bruker. *Rocky Mountain Nanotechnology Pt wire specifications.* URL: https://www.brukerafmprobes.com/p-3774-rmn-25pt300b.aspx.

[58] S. Mishra, A. K. Mondal, S. Pal, T. K. Das, E. Z. Smolinsky, G. Siligardi, and R. Naaman. "Length-dependent electron spin polarization in oligopeptides and DNA". In: *Journal of Physical Chemistry C* 124 (2020), pp. 10776–10782. DOI: 10.1021/acs.jpcc.0c02291.

[59] M. Kettner, B. Göhler, H. Zacharias, D. Mishra, V. Kiran, R. Naaman, C. Fontanesi, D. H. Waldeck, S. Sek, J. Pawowski, and J. Juhaniewicz. "Spin filtering in electron transport through chiral oligopeptides". In: *Journal of Physical Chemistry C* 119 (2015), pp. 14542–14547. DOI: 10.1021/jp509974z.

[60] Z. Xie, T. Z. Markus, S. R. Cohen, Z. Vager, R. Gutierrez, and R. Naaman. "Spin specific electron conduction through DNA oligomers". In: *Nano Letters* 11 (2011), pp. 4652–4655. DOI: 10.1021/nl2021637.

[61] C. H. Ko, Q. Zhu, F. Tassinari, G. Bullard, P. Zhang, D. N. Beratan, R. Naaman, and M. J. Therien. "Twisted molecular wires polarize spin currents at room temperature". In: *Proceedings of the National Academy of Sciences of the United States of America* 119 (2022), e2116180119. DOI: 10.1073/pnas.2116180119.

[62] P. Wagner, M. Hegner, H. J. Güntherodt, and G. Semenza. "Formation and in situ modification of monolayers chemisorbed on ultraflat template-stripped gold surfaces". In: *Langmuir* 11 (1995), pp. 3867–3875. DOI: 10.1021/la00010a043.

[63] T. Ederth. "Template-stripped gold surfaces with 0.4-nm rms roughness suitable for force measurements: Application to the Casimir force in the 20-100-nm range". In: *Phys. Rev. A* 62 (2000), p. 062104.

[64] J. Tepper. *Private Communication.* Leiden.

[65] C. Vericat, M. E. Vela, G. Benitez, P. Carro, and R. C. Salvarezza. "Self-assembled monolayers of thiols and dithiols on gold: New challenges for a well-known system". In: *Chemical Society Reviews* 39 (2010), pp. 1805–1834. DOI: 10.1039/b907301a.

[66] R. G. Nuzzo and D. L. Allara. "Adsorption of bifunctional organic disulfides on gold surfaces". In: *Journal of the American Chemical Society* 105 (1983), pp. 4481–4483.

[67] Y. Guo, M. Chen, and H. He. "Alkyl chain dependent alkanethiol self-assembled adsorption dynamics". In: *Surface Review and Letters* 22 (2015), p. 1550004. DOI: 10.1142/S0218625X15500043.

[68] J. C. Love, L. A. Estroff, J. K. Kriebel, R. G. Nuzzo, and G. M. Whitesides. "Self-assembled monolayers of thiolates on metals as a form of nanotechnology". In: *Chemical Reviews* 105 (2005), pp. 1103–1169. DOI: 10.1021/cr0300789.

[69] G. Yang and G.-Y. Liu. "New insights for self-assembled monolayers of organothiols on Au(111) revealed by scanning tunneling microscopy". In: *The Journal of Physical Chemistry B* 107 (2003), pp. 8746–8759. DOI: 10.1021/jp0219810.

[70] G. E. Poirier. "Characterization of organosulfur molecular monolayers on Au(111) using scanning tunneling microscopy". In: *Chemical Reviews* 97 (1997), pp. 1117–1127. DOI: 10.1021/cr960074m.

[71] M. Godin, P. J. Williams, V. Tabard-Cossa, O. Laroche, L. Y. Beaulieu, R. B. Lennox, and P. Grütter. "Surface stress, kinetics, and structure of alkanethiol self-assembled monolayers". In: *Langmuir* 20 (2004), pp. 7090–7096. DOI: 10.1021/la0302571.

[72] K. Edelmann, L. Gerhard, M. Winkler, L. Wilmes, V. Rai, M. Schumann, C. Kern, M. Meyer, M. Wegener, and W. Wulfhekel. "Light collection from a low-temperature scanning tunneling microscope using integrated mirror tips fabricated by direct laser writing". In: *Review of Scientific Instruments* 89 (2018), p. 123107. DOI: 10.1063/1.5053882.

[73] H. D. Wiederick, N. Gauthier, D. A. Campbell, and P. Rochon. "Magnetic braking: Simple theory and experiment". In: *American Journal of Physics* 55 (1987), pp. 500–503. DOI: 10.1119/1.15103.

[74] N. N. Greenwood and A. Earnshaw. *Chemistry of the Elements*. Butterworth-Heinemann, 1984.

[75] T. Bürgi. "Properties of the gold-sulphur interface: From self-assembled monolayers to clusters". In: *Nanoscale* 7 (2015), pp. 15553–15567. DOI: 10.1039/c5nr03497c.

[76] Y. Xue, X. Li, H. Li, and W. Zhang. "Quantifying thiol-gold interactions towards the efficient strength control". In: *Nature Communications* 5 (2014), p. 4348. DOI: 10.1038/ncomms5348.

[77] A. Maeland and T. B. Flanagan. "Lattice spacings of gold-palladium alloys". In: *Canadian Journal of Physics* 42 (1964), pp. 2364–2366.

[78] N. Takeuchi, C. T. Chan, and K. M. Ho. "Au(111): A theoretical study of the surface reconstruction and the surface electronic structure". In: *Physical Review B* 43 (1991), pp. 13899–13906. DOI: 10.1103/PhysRevB.43.13899.

[79] J. V. Barth, H. Brune, G. Ertl, and R. J. Behm. "Scanning tunneling microscopy observations on the reconstructed Au(111) surface". In: *Physical Review B* 42 (1990), p. 9307.

[80] L. Mougel. "Scanning Tunneling Microscopy study of tetraphenylmethane and s-triazine derivatives on Au (111) surface". Master thesis. Karlsruhe Institute of Technology, 2018.

[81] J. Homberg, M. Lindner, L. Gerhard, K. Edelmann, T. Frauhammer, Y. Nahas, M. Valášek, M. Mayor, and W. Wulfhekel. "Six state molecular revolver mounted on a rigid platform". In: *Nanoscale* 11 (2019), pp. 9015–9022. DOI: 10.1039/c9nr00259f.

[82] M. Lindner, M. Valášek, M. Mayor, T. Frauhammer, W. Wulfhekel, and L. Gerhard. "Molecular graph paper". In: *Angewandte Chemie - International Edition* 56 (2017), pp. 8290–8294. DOI: 10.1002/anie.201703586.

[83] T. Knaak, C. González, Y. J. Dappe, G. D. Harzmann, T. Brandl, M. Mayor, R. Berndt, and M. Gruber. "Fragmentation and distortion of terpyridine-based spin-crossover complexes on Au(111)". In: *Journal of Physical Chemistry C* 123 (2019), pp. 4178–4185. DOI: 10.1021/acs.jpcc.8b11242.

[84] L. Gerhard, K. Edelmann, J. Homberg, M. Valášek, S. G. Bahoosh, M. Lukas, F. Pauly, M. Mayor, and W. Wulfhekel. "An electrically actuated molecular toggle switch". In: *Nature Communications* 8 (2017), p. 14672. DOI: 10.1038/ncomms14672.

[85] M. Lindner, M. Valášek, J. Homberg, K. Edelmann, L. Gerhard, W. Wulfhekel, O. Fuhr, T. Wächter, M. Zharnikov, V. Kolivoška, L. Pospíšil, G. Mészáros, M. Hromadová, and M. Mayor. "Importance of the Anchor Group Position (Para versus Meta) in Tetraphenylmethane Tripods: Synthesis and Self-Assembly Features". In: *Chemistry - A European Journal* 22.37 (2016), pp. 13218–13235. DOI: 10.1002/chem.201602019.

[86] P. Markus. "Untersuchung von Tetrabiphenylmethanderivaten als molekulare Motoren mit Hilfe von Rastertunnelmikroskopie". Master Thesis. Karlsruher Institut für Technologie, 2021.

[87] X. Michalet. "Mean square displacement analysis of single-particle trajectories with localization error: Brownian motion in an isotropic medium". In: *Physical*

Review E - Statistical, Nonlinear, and Soft Matter Physics 82 (2010), p. 041914. DOI: 10.1103/PhysRevE.82.041914.

[88] V. N. Rai. "Light emission from single self-decoupled molecules in a scanning tunnelling microscope". PhD thesis. Karlsruhe Institute of Technology, 2021.

[89] J. B. Neaton, M. S. Hybertsen, and S. G. Louie. "Renormalization of molecular electronic levels at metal-molecule interfaces". In: *Physical Review Letters* 97 (2006), p. 216405. DOI: 10.1103/PhysRevLett.97.216405.

Acknowledgements

These past years have been some of the most rewarding, challenging and interesting times that I have experienced. I want to express my gratitude for the support of the people that guided me through these times.

First and foremost, I want to thank Prof. Dr. Wulfhekel for giving me the opportunity to work on this topic in his group. Thank you for your support and guidance throughout my PhD, be it scientifically, or personally. I admire you for how much you care about the people that surround you and the relaxed atmosphere that you create. I am very lucky to have you as the supervisor of my PhD.

I thank Prof. Dr. Wolfgang Wernsdorfer for agreeing to be the second referee of this thesis.

I am very grateful to Dr. Lukas Gerhard for all the guidance, the discussions and the help in and around the lab. You have helped me tremendously to come so far, not only with this thesis, but also in becoming a better person and scientist. I enjoyed the scientific and non-scientific talks and generally the time we have spent together. You set a great example for everyone in the group.

I would like to thank our long-time collaborators Dr. Nico Balzer, Dr. Michal Valášek and Prof. Dr. Marcel Mayor for synthesizing and providing us with the molecules used in this thesis, for discussing and accommodating to our needs.

My thanks are also with Dr. Jan Wilhelm and Prof. Dr. Ferdinand Evers for providing support from the theory side and for the calculations of the molecular structure and orbitals.

Moreover, I would like to express my gratitude to Prof. Dr. Jan M. van Ruitenbeek for welcoming me to his group for the duration of my Research Travel Grant at Leiden University, for helpful discussions and your calm way to approach problems. I thank the KHYS for the financial support of this stay.

Thank you, Federica, for teaching me the way of the AFM, for very helpful discussions and hints on experimental challenges. Thanks also go to Jędrzej Tepper for helping me during my stay in Leiden, be it through scientific and non-scientific discussion, using the SEM or working together. I'm very grateful to Norman, not only for the discussions and the patterning of the floppy disk, but also for bringing people together. It helped me a lot, to get to know so many nice people in Leiden, such as Peter, Jimi, Tycho, Guido, Alicja, Ceyhun and all the others. Personal thanks go to Gesa, it was a great pleasure to live with you in a crazy house with so many interesting people, to talk to you and to share the pleasure and pain of working in science.

Many thanks go to the people who helped me in the lab at Campus North and made life there more enjoyable: Paul, Melvin and especially Philipp, for your programming and measuring skills and your help with the new shields, and Kevin, for the smooth handing over of the 5K STM and for teaching me all the tricks with the machine. Special thanks go to Vibhuti, who was always happy to help out, discuss or just talk about everything and anything.

I would also like to express my gratitude to all the other members of the Wulfhekel group that I have had the pleasure to meet and spend time with over the years of PhD: Timo, Loïc, Thomas, Hongyan, Marie, Timofey, Fang, Hung-Hsiang, Gabriel, Anika, Simon, Mirjam, Qing, Avishek, Namrata, Jinjie and Qili. You all made life in the group fun and enjoyable.

My gratitude is also with my personal friends, without whom I would not be where I am now. Tim and Manu, we've known each other for so long, it is hard to imagine my life without you; Sebi, Martin, Mirko and Wieland without you, I would not have been able to even finish my bachelor's and I would have missed out on meeting some of the nicest people I've ever known; Julian, Devran and Mesel, Maria, all the people from H15 and so many others, thank you for being part of my life.

Last, but definitely not least, for the unwavering support throughout my whole life, in good and bad times and for always being there for me when I need it the most, I thank my family.

Experimental Condensed Matter Physics
(ISSN 2191-9925)

Band 1	Alexey Feofanov **Experiments on flux qubits with pi-shifters.** ISBN 978-3-86644-644-1
Band 2	Stefan Schmaus **Spintronics with individual metal-organic molecules.** ISBN 978-3-86644-649-6
Band 3	Marc Müller **Elektrischer Leitwert von magnetostriktiven Dy-Nanokontakten.** ISBN 978-3-86644-726-4
Band 4	Torben Peichl **Einfluss mechanischer Deformation auf atomare Tunnelsysteme – untersucht mit Josephson Phasen-Qubits.** ISBN 978-3-86644-837-7
Band 5	Dominik Stöffler **Herstellung dünner metallischer Brücken durch Elektromigration und Charakterisierung mit Rastersondentechniken.** ISBN 978-3-86644-843-8
Band 6	Tihomir Tomanic **Untersuchung des elektronischen Oberflächenzustands von Ag-Inseln auf supraleitendem Niob (110).** ISBN 978-3-86644-898-8
Band 7	Lukas Gerhard **Magnetoelectric coupling at metal surfaces.** ISBN 978-3-7315-0063-6
Band 8	Nicht erschienen.
Band 9	Jochen Zimmer **Cooper pair transport in arrays of Josephson junctions.** ISBN 978-3-7315-0130-5
Band 10	Oliver Berg **Elektrischer Transport durch Nanokontakte von Selten-Erd-Metallen.** ISBN 978-3-7315-0209-8
Band 11	Grigorij Jur'evic Grabovskij **Investigation of coherent microscopic defects inside the tunneling barrier of a Josephson junction.** ISBN 978-3-7315-0210-4

Die Bände sind unter www.ksp.kit.edu als PDF frei verfügbar oder als Druckausgabe bestellbar.

Band 12 Cornelius Thiele
STM Characterization of Phenylene-Ethynylene Oligomers on Au(111) and their Integration into Carbon Nanotube Nanogaps.
ISBN 978-3-7315-0235-7

Band 13 Michael Peter Schackert
Scanning Tunneling Spectroscopy on Electron-Boson Interactions in Superconductors.
ISBN 978-3-7315-0238-8

Band 14 Susanne Butz
One-Dimensional Tunable Josephson Metamaterials.
ISBN 978-3-7315-0271-5

Band 15 Philipp Jung
Nonlinear Effects in Superconducting Quantum Interference Meta-Atoms.
ISBN 978-3-7315-0294-4

Band 16 Sebastian Probst
Hybrid quantum system based on rare earth doped crystals.
ISBN 978-3-7315-0345-3

Band 17 Wolfram Kittler
Magnetische Anisotropie und Quantenphasenübergang in $CeTi_{1-x}V_xGe_3$.
ISBN 978-3-7315-0363-7

Band 18 Moritz Peter
Towards magnetic resonance in scanning tunneling microscopy using heterodyne detection.
ISBN 978-3-7315-0410-8

Band 19 Junji Tobias Märkl
Investigation of Magnetic Adatoms with Scanning Tunneling Techniques.
ISBN 978-3-7315-0435-1

Band 20 Jochen Braumüller
Quantum simulation experiments with superconducting circuits.
ISBN 978-3-7315-0780-2

Band 21 Jinjie Chen
Local Investigation of Single Magnetic Molecules with Scanning Tunneling Microscopy.
ISBN 978-3-7315-0819-9

Band 22 Arnold Seiler
Einfluss der Leitungselektronen auf die Dynamik atomarer Tunnelsysteme in ungeordneten Festkörpern: Relaxationsprozesse in metallischen Gläsern und ungeordneten dünnen Aluminiumoxid-Schichten.
ISBN 978-3-7315-0870-0

Band 23 Jasmin Maria Jandke
Elastic and Inelastic Scanning Tunneling Spectroscopy on Iron-Based Superconductors.
ISBN 978-3-7315-0747-5

Band 24 Kevin Edelmann
Electroluminescence from Plasmonic Excitations in a Scanning Tunnelling Microscope.
ISBN 978-3-7315-0923-3

Band 25 Lei Zhang
Sub-Kelvin scanning tunneling microscopy on magnetic molecules.
ISBN 978-3-86644-950-3

Band 26 Lukas Grünhaupt
Granular aluminium superinductors.
ISBN 978-3-7315-0956-1

Band 27 Alexander Bilmes
Resolving locations of defects in superconducting transmon qubits.
ISBN 978-3-7315-0967-7

Band 28 Marco Pfirrmann
Adding nonlinearity to an electromagnetic-magnonic quantum hybrid device.
ISBN 978-3-7315-1003-1

Band 29 Andre Schneider
Quantum Sensing Experiments with Superconducting Qubits.
ISBN 978-3-7315-1032-1

Band 30 Vibhuti Narayan Rai
Light Emission from Single Self-decoupled Molecules in a Scanning Tunnelling Microscope.
ISBN 978-3-7315-1201-1

Band 31 Julian Antoni Skolaut
Molecular Motor Based on Single Chiral Tripodal Molecules Studied with STM.
ISBN 978-3-7315-1247-9
